Nuclear Structure
in China 2010

Nuclear Structure in China 2010

Proceedings of the 13th National Conference on
Nuclear Structure in China

Chi-Feng, Inner Mongolia, China, 24 – 30 July 2010

editors

Hong-Bo BAI
Chi-Feng University, China

Jie MENG
Peking University, China & Beihang University, China

En-Guang ZHAO
Institute of Theoretical Physics, Chinese Academy of Sciences, China

Shan-Gui ZHOU
Institute of Theoretical Physics, Chinese Academy of Sciences, China

World Scientific

NEW JERSEY · LONDON · SINGAPORE · BEIJING · SHANGHAI · HONG KONG · TAIPEI · CHENNAI

Published by

World Scientific Publishing Co. Pte. Ltd.

5 Toh Tuck Link, Singapore 596224

USA office: 27 Warren Street, Suite 401-402, Hackensack, NJ 07601

UK office: 57 Shelton Street, Covent Garden, London WC2H 9HE

British Library Cataloguing-in-Publication Data
A catalogue record for this book is available from the British Library.

NUCLEAR STRUCTURE IN CHINA 2010
Proceedings of the 13th National Conference on Nuclear Structure in China

Copyright © 2011 by World Scientific Publishing Co. Pte. Ltd.

ISBN-13 978-981-4360-64-7
ISBN-10 981-4360-64-3

Printed in Singapore.

PREFACE

The 13th National Conference on Nuclear Structure in China (NSC2010) was held from July 25 to 31, 2010 in Chi-Feng city, Inner Mongolia and hosted by Chi-Feng University. The series of National Conferences on Nuclear Structure in China is organized by the Nuclear Structure Sub-committee of Chinese Nuclear Physics Society and it is held biennially. This series of conferences is an important event in China for nuclear physicists working on the study of nuclear structure.

Apart from the Nuclear Structure Sub-committee of Chinese Nuclear Physics Society, NSC2010 was also organized by China Center of Advanced Science and Technology (CCAST), Peking University, Institute of Theoretical Physics of Chinese Academy of Sciences, Institute of Modern Physics of Chinese Academy of Sciences, China Institute of Atomic Energy, and Chi-Feng University. There were about 160 participants from more than 40 universities or institutes, including some invited speakers from abroad, e.g., Prof. R. J. Liotta from KTH, Sweden.

Following the tradition of NSC conferences, NSC2010 presented a review of recent advances made in nuclear structure physics in China and provided a broad discussion forum on current and future research projects. The main focus was on the following topics: exotic nuclear structure, structure and synthesis mechanism of superheavy nuclei, nuclear high-spin states, nuclear magnetic rotation and chirality, nuclear astrophysics, nuclear matter and symmetry energy, hypernuclear physics, and nuclear many body approaches.

In each NSC conference, the students and junior researchers make up half or more than half of the participants. Starting from the 11th National Conference on Nuclear Structure in China (NSC2006) which was held in Guiyang, Guizhou province, best oral presentations were selected from the students and junior researchers by the selection committee which is headed by Prof. En-Guang Zhao (赵恩广) and Prof. Yong-Shou Chen (陈永寿). In NSC2010, best oral presentations included: Hao-Zhao Liang (梁豪兆) (Peking University), Zhong-Min Niu (牛中明) (Peking University), Yue Shi (石跃) (Peking University), Xiao-Yan Sun (孙小燕) (Shanghai Institute of Applied Physics, Chinese Academy of Sciences), Jian-Guo Wang (王建国) (Tsinghua University), Shuo

Wang（王硕）(Peking University), Yong-Jia Wang（王永佳）(Lanzhou University), Yun Zheng (郑云) (China Institute of Atomic Energy), and Zhen-Hua Zhang (张振华) (Institute of Theoretical Physics, Chinese Academy of Sciences).

We thank the members of the Nuclear Structure Sub-committee of Chinese Nuclear Physics Society and the members of Local Organizing Committee for their support and valuable advices on the organization of NSC2010. We would like to take this opportunity to extend our sincere gratitude to all participants for their attending, contributions and discussions. We also thank Mr. Kai Wen (温凯) and Mr. Zhen-Hua Zhang (张振华) for their help in preparing this proceeding. Finally, the generous financial supports of the organizing institutions as well as the excellent and efficient work of World Scientific Publishing Co. are gratefully acknowledged.

Hong-Bo Bai (白洪波)	*Chi-Feng University, Chi-Feng*
Jie Meng (孟杰)	*Peking University &* *Beihang University, Beijing*
En-Guang Zhao (赵恩广)	*Institute of Theoretical Physics,* *Chinese Academy of Sciences, Beijing*
Shan-Gui Zhou (周善贵)	*Institute of Theoretical Physics,* *Chinese Academy of Sciences, Beijing*

CONTENTS

PERSONAL VIEW ON NUCLEAR PHYSICS RESEARCH
关于核物理研究的一点思考

JIE MENG

State Key Laboratory of Nuclear Physics and Technology, School of Physics, Peking University, Beijing 100871, China
School of Physics and Nuclear Energy Engineering, Beihang University, Beijing 100191, China

This article is based on the opening speech given at the thirteenth Chinese biennial Nuclear Structure Conference held in Chifeng, Inner Mongolia, China. The history of this series of Chinese biennial nuclear structure conference is briefly introduced. Personal views on nuclear physics research are presented following the overview of the opportunity provided by main nuclear physics facilities world-wide, important sub-field in nuclear physics, main achievement including important progress in Chinese nuclear physics community, the related important series of international nuclear physics conferences, population of nuclear physicists, newly emerging group in nuclear physics and possible improvement on organizing the Chinese biennial nuclear structure conference.

文章简要介绍了历次全国核结构大会的历史,并通过对部分国内外大型核物理研究设备现状、当前核物理的重要研究领域、国内外核物理的几个重要进展、国际上重要的核物理会议、核物理领域科普宣传、国内核物理新兴研究基地的介绍,对核物理研究以及如何办好核结构大会等提出一些思考和看法。

作为核结构专业委员会主任,上一次在全国核结构大会作大会报告,应该在 2006 年。那次会议是受核结构专业委员会的委托,特别地也是曾谨言先生的提议,让我介绍一下核结构发展的几个里程碑以及中国核结构研究 20 年的成就[1]。

自从那次会议之后,核结构专业委员会的几位同事也一直在思考怎样将核结构大会办得更好。大家达成一个基本共识,就是在核结构大会期间,应该委托主任或者副主任做一个工作报告,这个报告的内容可能很多也很杂。我今天的报告就是这个目的,题目为"Personal view on nuclear physics research",很多是个人看法,也不一定很成熟。

报告的主要内容包括:引言、2008 年以来核物理中几个重要的工作、中国的部分核物理工作、重要的国际核物理会议包括 INPC2010(核核碰撞会议由于在中国召开,参会的人很多,这里就不用介绍了)。接下来给大家展示一些人才培养数据、学界新人（指这几年核物理专业毕业的博士，并在国内的研究院所或高校取得固定位置的年轻人）。最后就如何办好核结构大会提出一些建议，希望得到大家指正。

1. 引言

第一届全国核结构大会始于 1986 年，距今四分之一世纪。这个会议的主要发起人是曾谨言先生和孙洪洲先生，承办人是西南师范大学（现在的西南大学）殷传宗教授和林辛未教授，当时殷传宗教授任西南师范大学物理系系主任。这个会议的一个重要的决定就是这样的会议还得继续下去，并委托中国科学院理论物理研究所赵恩广研究员作为联系人。

第二届全国核结构大会于 1988 年在安徽合肥召开，很遗憾我们缺少一张集体照。第三届全国核结构大会于 1990 年在大连召开，徐躬耦先生也参会了。这次会议的承办人是喻身启教授，潘峰教授做了许多具体工作。潘峰教授现任辽宁师范大学物理学院院长，他也是赤峰人，所以第十三届全国核结构大会到这儿来开，我们每距几年总要到原来开会地点转一转。从第四届全国核结构大会起，我暂时脱离组织。2002 年在合肥开会的时候我说过，第一次就参会的人员现在还在的已经不多了，我是其中之一。

第四届全国核结构大会在湖南张家界召开，参会的人数比较多。这次会议有一个里程碑性质的作用，它标志着全国核结构大会参会人数突破一百人。第五届全国核结构大会是在湖北宜昌召开的，与会人数也不少，这可能与当时的三峡告别游有关。第六届全国核结构大会由近代物理所举办，规模非常大，人数也非常多。第七届全国核结构大会在银川召开，大家可以发现这次会议人数一下子就少了。所以在这个时候，大家酝酿着核结构大会一定要开得热闹些，所以才有了规模非常大的第八届全国核结构大会。第八届全国核结构大会在海南海口召开，有丁大钊、张焕乔、叶铭汉三位院士与会，同时还有现在的原子能院院长赵志祥，当时的兰州近物所的所长罗亦孝，副所长靳根明等强大阵容。

第九届全国核结构大会在安徽大学召开，因为安徽大学也是我们的老根据地，同时也是因为徐辅新教授在他临退休前要为核物理贡献一点力量。这次会议之后还做出了一个大胆的决定，就是赵恩广先生决定把组织这个会议的任务交给我。但是，实际上他一直都在做着大量烦琐、实际的组织工作。

这次会议之后，2004 年的第十届全国核结构大会在贵阳召开。2006 年第十一届全国核结构大会在吉林大学召开，我不能确认这是否是吴式枢先生最后一次正式出席的大会，此外吉林大学学术委员会主任邹广田院士也出席了会议。第十二届全国核结构大会我们又回到了重庆，这次会议一共有 30 个单位参加，与会代表是 140 多人。

2. 核物理研究装置

这段时间我在思考一个问题，就是核物理怎么发展？怎么像粒子物理、凝聚态、天体物理等学科一样能够证明它的独特性及其存在的特殊价值？为什么会想到这个问题呢？大家都知道不管是从卢瑟福还是从居里夫人还是贝克勒尔开始，核物理已经走过了一百多年历程。那么这一百多年之后的我们还要研究什么？解决什么问题？这是一个值得我们深思的问题。这几年我也搜集了一些材料和题材供大家参考，其中包括参加国外会议如 NUSTAR2010 和 INPC2010 以及在俄罗斯参加的几次会议搜集的材料。

大家知道大型核物理研究设备的建设这几年方兴未艾。不久前有一个报道：物理学家在几天之内就发现了几十种新核素，这实际上就是日本理化学研究所新一代放射性束装置 RIBF 做出的成果之一。同时，J-PARC 去年正式建成，全世界都欢欣鼓舞。李政道先生在给 Nagamiya 祝贺的时候提到，希望 J-PARC 除了产生强流质子之外，还有一个功能，就是从此之后能够不断加速诺贝尔奖获得者的产生。大家知道 FAIR project，德国很早就开始筹划了。它要建成像 CERN 一样的国际合作装置。今年三月份在德国的 NUSTAR 会议上，FAIR 的时间表进行了更新。依照新的时间表，如果每一步都按照计划完成，那么整个项目大概在 2017 年能完成。如果 2017 年才能出物理结果的话，我会很高兴，因为至少我退休之前，核物理的工作还是做不完的。欧盟合作提出的新放射性束装置 EURISOL 按照时间表计划是 2026 年建成。大家知道，美国在 MSU 建造的 FRIB 耗资约 5 亿美金。在国际上很热的 KoRIA 和韩国的称呼非常相近。KoRIA 实际上是 Korea Rare Isotope Accelerator 的缩写，是韩国根据美国 FRIB 引进消化吸收的放射性束设备，他们的经费已经全部到位，这给世界核物理带来很大的震撼。因为韩国核物理的基础薄弱，但是即使像韩国没有核物理的基础，通过引进消化吸收后可以出口反应堆到中东一样，韩国未来的发展也是不可限量。美国、日本、欧洲等都积极参与韩国的 KoRIA 建造。

近几年来核物理的重要研究领域包括什么呢？当然有非稳定原子核的研究，这是一个方兴未艾的领域。其次是核天体物理、标准模型检验、超子物理以及 QCD 物理等等，这些都是核物理比较时髦的领域。

还有一个问题就是关于核物理的定义。为什么要提到核物理的定义呢？这是几年在国外开会的时候和国外同行交流的主要话题之一。欧洲的核物理这几年的发展势头和日本相比减缓了很多，其中很重要的一个原因就与核物理的定义有很大关系。在欧洲，除了找不到原因来换用别的名字如亚原子物理、有限费米子体系等，几乎很少有人主动提及自己是做核物理的。主要的一个原因就是要回避"核物理"这个名词。记得 2000 年在克罗地亚参加 NATO 资助的一个核物理的 workshop 时，一个德国教授告诉我为什么他们要回避核物理这个名词，理由是前苏联的切尔诺贝利核泄漏使得欧洲有恐核心理。当时，我举了一个例子告诉他，我说我们中国现在蓬勃兴起的是汽车工业，交通事故不断增加，但是从来不会把这个原因归咎于奔驰或福特，那肯定是驾驶员的失误造成的。同样，前苏联的切尔诺贝利核泄漏也应该是管理者的责任，而不应该是科学家的责任，核物理本身是没有责任的。

日本的情况却恰恰与之相反，五年前欧洲很多科学家就认为日本即将超越他们。这个月初在北美开会的时候，很多美国人和加拿大人都认为现在领导世界核物理发展的龙头是日本。在日本，核物理的范畴很宽，从核物理延伸出来的许多领域依然属于核物理，它没有中高能物理或者强子物理等分支，不同领域之间相互交叉很多，这也许是日本核物理发展很快的一个原因。日本核物理的影响还有一个例子，每隔几年，日本和美国联合在 Hawaii 召开核物理年会，与会代表数接近千人。记得在 911 事件之后，日本和美国联合在 Hawaii 召开首次核物理年会，记者问日本核物理年会组织主席，日本和美国在一起开年会代表什么。日方主席说这标志着日本的核物理学界而不仅仅是个别核物理学家完全被美国同行所接受。希望有一天，中国在与日本和美国开核物理年会的时候，同样可以自豪地宣布，这标志中国的核物理界完全被世界所接受。这个所谓的接受，就意味着我们真正进入世界一流。

3. 核物理的几个重要进展

以上内容说明核物理目前仍然是一个重要的前沿科学领域。Physical Review C 2009 年的数据可以说明核物理研究的现状。核物理至少包括核子-核子相互作用、少体系统、核结构、核反应、相对论核碰撞、强子物理与 QCD、标准模型对称性和核天体物理等领域。

重要的核物理杂志 Physic Review C 影响因子 2009 年是 3.477，每年发表的文章是 1087 篇，共 9805 页[2]。Physical Review C 2009 年的数据同时还给出了每个国家投稿的总数。美国的投稿数量和发表数量都在呈现相对下降趋势，但是欧洲在增加。很遗憾的是到现在为止，中国还被统计在亚太地区而没有被单列出来。

美国物理学会网站 http://physics.aps.org/，列出了一些 viewpoint on nuclear physics。从 2008 年到 2010 年，核物理有 15 篇文章，Synopses 有 19 篇文章，共 34 篇文章。这些文章列出来，可以展现这几年核物理的主要进展。

首先，一个重要的进展是 2010 年 7 月 8 日在 Nature 上发表的有关 proton size 文章[3]。实验测量出来的质子半径和过去的结果相差了 4%。如果类比其它一些领域的发展，会发现所有热门领域的发展都经历了这样一个过程：实验与理论相差 4% 就可以称为 puzzle，或者 abnormal，这是一个领域兴起的信号；接下来就是 crisis，如 spin crisis（质子自旋危机）；有了 puzzle 和 crisis 之后，这个领域就会有很大的发展。这是 Nature 很少有的以核物理研究成果作为封面的文章，可以算核物理的一件大事。

第二个重要进展应该是 117 号元素的合成[4]。这是俄罗斯 Dubna 联合核研究所几十年奋斗的结果。它标志着我们在核素图上艰难的探索又迈了非常重要的一步。在 INPC2010 会议上，美国国家能源部的 D. Dean 介绍核物理研究的美国国家政策，列举了许多核物理推动人类进步和发展的例子，合成 117 号元素就列在其中，同时强调 117 号元素的合成也有美国劳伦斯利弗莫尔国家实验室、美国橡树岭国家实验室、范德堡大学、内华达大学的贡献。

在不稳定核方面取得的一个重大突破是 ^{22}C 的合成[5]，这也是核物理的一个重要发现。另外一个工作是理论计算，通过液滴模型计算，把核物质的状态方程推向了低密度的情况[6]。其实核物理研究发展到今天，我为什么坚信核物理的发展空间还很大呢？因为大家都知道，我们的世界 99% 以上的质量来自原子核。但实验给出的原子核性质基本上就是饱和点附近的性质，比这个密度高或者低的核物质性质都不能精确测量。

He $+^{208}$Pb 实验也是一个重要的工作[7]，它通过 E1 跃迁证明或间接证明了 α 粒子在原子核中是存在的。

张量力是近年来国际上一个重要的研究热点，张量力究竟有多大的贡献，其实我们并不清楚。但是这方面，日本同行把张量力用到壳模型中，可以再现壳层结构的同位旋依赖性，并指出这是张量力存在的证据[8]。

还有一个特别重要的工作是原子核有球形和形变形状共存[9]，它与原子核高自旋研究密切相关。过去几十年高自旋态的研究中，我们认为轻核没有集体转动性，重核会发生裂变。而在 ^{258}Fm 发现的高自旋转动带[10]，对我们过去的信念产生一个很大的冲击，表明重的原子核里面同样是可以有高自旋态的。

下一个重要的工作是关于核反应截面的测量[11]，它的重要性在于核反应截面的测量与星系里的热核过程相关。

牛中明等人的工作也引起很大关注[12]，该工作把核物理的知识运用到天体元素合成的计算过程中。被美国物理学会的编辑起了一个名字: Calibrating the cosmic clock——为宇宙的时钟进行标度。

最近在 Physics Review Letters 还有一个工作，就是 Three-Body Forces and the Limit of Oxygen Isotopes[13]。大家知道到目前为止，通常的核结构模型中，^{28}O 是束缚的。最近，研究壳模型的科学家引入三体力可以把 ^{28}O 变成非束缚。同样，这个工作和张量力的工作有些类似。但是，我关心如果对 F 和 Ne 进行将类似计算，最后的结果会是怎样的？

还有很多重要工作，由于时间限制不再做详细介绍。同时在过去两年中，中国的核物理有很多工作值得骄傲，这里就简单的举几个例子。

关于 PRL 的文章，2008 年我们有两篇文章：一篇是梁豪兆等人关于 Spin-Isospin Resonances: A Self-Consistent Covariant Description 的文章[14]；另外一篇付伟杰等人关于中子星方面的文章[15]。2009 年有 N.

Paar 和牛一斐等人的工作[16]，以及亓冲等人关于集团衰变的工作[17]。应该特别指出的是上周刚被接收的白春林等人的文章[18]。白春林是张焕乔先生的博士，今年刚取得博士学位，现在在四川大学工作，他们研究张量力对原子核 ^{208}Pb 的共振态影响。

中国还有许多工作在很多国际上重要文章中被引用。发表在 PRL 上的 117 号元素实验合成文章中[4]，引用了国内两篇理论的文章。其中一篇文章是沈彩万等人发表在 International Journal of Modern Physics E 的文章[19]。可以说，任何时候，只要有好工作，在任何杂志上发表都是可以让大家知道的。我也是 International Journal of Modern Physics E 的 Managing editor，这个杂志还接收一些 review 文章，所以以后大家有感兴趣的题目可以告诉我。另外一篇文章是刘祖华等人的文章[20]。今年 6 月底在俄罗斯开会时，白春林等人发表在 Physics Letter B 和 Physics Review C 的工作[21]被俄罗斯人大量引用和评述。李志攀等人关于量子相变的工作[22]，在去年的欧洲核物理年会和土耳其召开的对称性的会议上被 F. Iachello 和 R. F. Casten 引用。另外，梁豪兆和尧江明的工作[23,24]在 INPC2010 上被引用。

4. 国际上重要的核物理会议

俄罗斯 Dubna 每三年召开一次的核结构大会－－Structure and Related Topics—NSRT，从上个世纪 60 年代开始，持续了差不多半个世纪，建议大家关注。同时，下个月在美国 Berkeley 还有一个在北美召开的核结构系列大会，这些都是非常好的会议。

借此机会，给大家介绍一下 INPC2010 会议的情况。整个会议的与会人员约 750 人，注册人员约 830 人，其中有约 130 个学生。中国参加的人数约 10 人，所以我们的比例相当相当小。我在参加会议时遇到两个知名科学家：一个是 R. Tribble, 现任日本理化学研究所束流委员会主任。这次会议给他安排了一个展板，他就站在那里认认真真、一丝不苟的给大家解释他的展板。另一个是 J. H. Hamilton，他与清华大学朱胜江教授有很好的合作，也获得过中国的国家科技合作的友谊奖。同样的，他也在那儿做一个展板。我很敬佩他们这种积极参与的态度。

会议的开幕致辞是 GSI 的 P. Braun 做的，他主要是讲核物理的发展。会议有 10 个大会模块和 40 个大会邀请报告，包括 Hot and dense QCD, Hadrons in nuclei, Nuclear astrophysics, Nuclear Reactions, New facilities and instrumentation, Applications and interdisciplinary research, Hadron structure, Nuclear structure, Standard model tests and fundamental symmetries, Neutrinos and nuclei.

在 INPC2010 会议上，日本有四个大会报告，印度有一个大会报告。俄罗斯有一个大会报告，是 Yu. Oganessian 讲的 117 号元素。大会主席里有两个来自日本。在大会报告和大会主席里面没有安排中国人，大家觉得有点遗憾。同样，日本同行也在抱怨，整个大会报告里面没有安排日本理化学研究所的人。去看一下整个会议的日程，就会发现安排的加拿大会议主席和大会报告还是较多的。所以，从这个角度上看，我们在国内主办国际会议时，怎么安排和突出中国人的工作，同样值得思考。

分会场一共有 48 个，有 76 个邀请报告，有 185 个口头报告。中国有 5 个口头报告：周善贵、梁豪兆、柳卫平、蔡翔舟、方德清；一个分会主席：柳卫平。学生的模块就是展板，所有参会学生必须提供展板，整个会议结束后评选出一个优秀报告奖及 3 个展板奖。梁豪兆和牛一斐都进入了最后的小名单，但最后评出的 3 个展板奖和一个口头报告，3 个来自北美，而且其中多数还来自 TRIUMF。

大会报告中，Yu. Oganessian 的大会报告做的非常漂亮，因为它让任何与会的人，不管在不在这个领域的都能听得懂。核结构大会安排了四个邀请报告：第一个报告是 A. Gade 做的，她在美国阿贡国家实验室工作，所以邀请她做 γ 谱学的报告。第二个报告是德国的 A. Schwenk 做的，他是原 TRIUMF 理论部的主任。他做有效模型的大会报告，基本是介绍他和 T. Otsuka 等人的工作。第三个报告是 K. Blaum 做的，他是在 Heidenberg 做质量测量的，今年在德国获得 GENCO 奖。他的报告让我明白了一点，无论理论还是实验物理工作，只要你清楚问题在哪里，即使困难再大，一步一个脚印，同样可以做很好的工作。另外的大会报告是 J. Dobaczewiski 做的，他 2001 年在 INPC 作过一次大会报告，是世界上少有的两次做大会邀请报告的人。

在大会中也提到了很多中国的工作，其中包括郭冰和申虹的工作，同样也提到了中国的设备 CSR、BRIF，以及中国在建和设计正在申请的项目。当然，同样受到关注的还有韩国的 KoRIA。过去我们没有超过日本，希望未来我们不会担心赶不上韩国。

在强子结构领域中，孟大中、梁作堂、季向东等人的工作也被多次提及。A. Gade 在 γ 谱学的实验工作中引用了孙扬的一个理论工作。尧江明等人的工作[24]在 J. Dobaczewiski 的大会报告里有提及。梁豪兆等人的工作[23]在模型检验里面的大会报告被重点引用，认为引起了标准模型中 CKM 矩阵的 unitarity lost。所谓 unitarity lost 和我们前面说到的 puzzle 差不多一样。

和大家一样，近来我也在思考一个问题：Nuclear Physics in China, to be or not to be? 为此，我要感谢李璐璐博士帮我做了一个很重要的工作，就是收集整理我们能收集到的关于国内几个单位最近这两年以来，产出了多少博士，这些博士都在干什么的数据。因为原子能院的数据以及很多大学的数据没有统计，所以这只是一个初步统计。

2008 年到 2010 年，我们培养的博士人数大于 80，这是一个令人非常高兴的数字。同样的，我更关心学界的新人，他们在国内什么地方扎下根来。同样很遗憾，这里面也没有原子能院的数据。我非常高兴的看到我们 08 年以来至少有 40 个新人，如果按照这样的速度发展下去，中国的核物理是非常有希望的。这个数据非常不全面，如果大家有新的数据欢迎提供。

我关注的另一个问题就是核物理领域对外宣传。在这方面《一万个科学难题》和《Journal of Physics G: Nuclear and Particle Physics Volume》最近都做了一些努力[25,26]。

此外，在学科建设方面，存在一个有意思的现象，那就是在很多小学校有一个大的学科存在。我想举的两个例子是赤峰学院和湖州师范学院。我不知道这两个学校什么时候会提出国内一流或是国际一流的口号，但是这两个学校在核物理方面都做了很好的工作。另外一方面就是要重视年轻人的培养。这次 INPC2010 会议做得不错，年轻的学生去参加会议，不仅可以申请资助，而且会议期间给学生提供两次免费午餐：一次是让他们学习怎么管理好自己的学习和工作；另一次是告诉他们怎么做报告。我觉得这是值得我们借鉴的。

5. 关于核结构大会的一些思考

诸位，全国核结构大会到今年已经走过了四分之一个世纪，但是我们还是要想想这个核结构大会究竟应该怎么开，就像我们核物理往哪走一样。我们还是要强调不断要有新人出来，然后要扩大交流。如果大家有很好的建议，我们可以再修改、再提高。

我们可否形成这样一个固定的模式：大会的报告是 30+10 分钟，30 分钟的报告包括近年来该领域的总体情况介绍，再加上 10 分钟的讨论。大会报告中的典型工作报告是 15+5 分钟，这几年中国的确有一些很好的工作能代表中国的形象，我们就把它控制在 15+5 分钟这个模式。此外，就是其它的一些邀请报告，大家觉得有非常强、非常好的人到中国访问，可以邀请他们做一个大会报告，同样也是 30+10 分钟。分会报告是 20+5 分钟或 12+3 分钟。

最后，评出不多于 5 个优秀青年报告奖，然后颁发证书。我们希望让得奖人为在这个领域工作而感到骄傲和自豪，同时让其他人觉得这是一种鼓励和鞭策。现在评奖委员会主任是赵恩广先生，副主任是陈永寿先生，委员是由主任和副主任聘请的。评奖程序是本人申请（按会议组委会给出的时间节点，提交电子版个人简历、发表文章目录及 2 篇代表性论文）；评奖委员会参加申请人的报告，根据申请人准备报告、讲述和回答问题等环节的表现，综合评定，最后经评奖委员会讨论产生。在会议结束之前，我们颁发优秀青年报告奖。这次会议中，每个部分都安排了两个主席，目的一方面为了让大家共同把会议主持好，另一方面也是为了培养人才。在核物理领域里面除了工作以外，主席也采取老带新的形式，让更多的人承担更大的责任。

谢谢大家！

6. 致谢

感谢周善贵、李璐璐、陈颖在准备和整理报告过程中的帮助，同时也感谢国家自然科学基金（批准号：10720003, 10775004, 10435010, 10221003）、国家重点基础研究发展计划（批准号：2007CB815000）的资助。

References

1. 中国核物理学会核结构专业委员会, 高能物理与核物理 **30**, Supp. II:1-13 (2006).
2. http://prc.aps.org/about.
3. R. Pohl, A. Antognini, F. Nez, F. D. Amaro, F. Biraben *et al.*, *NATURE* **466**, 213 (2010).
4. Y. T. Oganessian, F. S. Abdullin, P. D. Bailey, D. E. Benker *et al.*, *Phys. Rev. Lett.* **104**, 142502 (2010).
5. K. Tanaka, T. Yamaguchi, T. Suzuki, T. Ohtsubo, M. Fukuda, D. Nishimura, M. Takechi *et al.*, *Phys. Rev. Lett.* **104**, 062701 (2010).
6. J. B. Natowitz, G. Röpke, S. Typel, D. Blaschke *et al.*, *Phys. Rev. Lett.* **104**, 202501 (2010).
7. A. Astier, P. Petkov, M.-G. Porquet, D. S. Delion and P. Schuck, *Phys. Rev. Lett.* **104**, 042701 (2010).
8. T. Otsuka, T. Suzuki, M. Honma, Y. Utsuno, N. Tsunoda *et al.*, *Phys. Rev. Lett.* **104**, 012501 (2010).
9. L. Gaudefroy, J. M. Daugas, M. Hass, S. Grévy *et al.*, *Phys. Rev. Lett.* **102**, 092501 (2009).
10. P. T. Greenlees, R.-D. Herzberg, S. Ketelhut, P. A. Butler *et al.*, *Phys. Rev. C* **78**, 021303 (2008).
11. C. Wrede, J. A. Caggiano, J. A. Clark, C. M. Deibel, A. Parikh and P. D. Parker, *Phys. Rev. C* **79**, 045803 (2009).
12. Z. Niu, B. Sun and J. Meng, *Phys. Rev. C* **80**, 065806 (2009).
13. T. Otsuka, T. Suzuki, J. D. Holt, A. Schwenk and Y. Akaishi, *Phys. Rev. Lett.* **105**, 032501 (2010).
14. H. Liang, N. Van Giai and J. Meng, *Phys. Rev. Lett.* **101**, 122502 (2008).
15. W. J. Fu, H. Q. Wei and Y. X. Liu, *Phys. Rev. Lett.* **101**, 181102 (2008).
16. N. Paar, Y. F. Niu, D. Vretenar and J. Meng, *Phys. Rev. Lett.* **103**, 032502 (2009).
17. C. Qi, F. R. Xu, R. J. Liotta and R. Wyss, *Phys. Rev. Lett.* **103**, 072501 (2009).
18. C. L. Bai, H. Q. Zhang, H. Sagawa, X. Z. Zhang, G. Colò and F. R. Xu, *Phys. Rev. Lett.* **105**, 072501 (2010).
19. C. W. Shen *et al.*, *Int. J. Mod. Phys. E* **17**, 66 (2008).
20. Z. H. Liu and J. D. Bao, *Phys. Rev. C* **80**, 034601 (2009).
21. C. Bai, H. Sagawa, H. Zhang, X. Zhang, G. Colò and F. R. Xu, *Physics Letters B* **675**, 28 (2009).
22. Z. P. Li, T. Nikšić, D. Vretenar, J. Meng, G. A. Lalazissis and P. Ring, *Phys.*

Rev. C **79**, 054301 (2009).

23. H. Liang, N. V. Giai and J. Meng, *Phys. Rev. C* **79**, 064316 (2009).
24. J. M. Yao, J. Meng, P. Ring and D. Vretenar, *Phys. Rev. C* **81**, 044311 (2010).
25. "10000个科学难题"物理学编委会, 10000个科学难题:物理学卷(精装) 科学出版社, (2009).
26. J. Dobaczewski *et al.*, *J. Phys. G: Nuclear and Particle Physics* **37**, 060301 (2010).

HIGH-SPIN LEVEL STRUCTURES IN ^{89}Zr

X. P. CAO

China Institute of Atomic Energy, Beijing 102413, China
Department of Physics, Northeast Normal University,
Changchun 130024, China

X. G. WU, C. Y. HE, L. H. ZHU, Z. M. WANG, Y. ZHENG, S. H. YAO, G. S. LI

China Institute of Atomic Energy, Beijing 102413, China
wxg@ciae.ac.cn
chuangye.he@gmail.com

X. Z. CUI

Department of Physics, Jilin University, Changchun 130023, China

The high-spin states in ^{89}Zr have been studied by in-beam γ-ray spectroscopy using the heavy ion fusion evaporation reaction ^{76}Ge(^{19}F,p5n)^{89}Zr reaction at a bombarding energy of 80 MeV. The level scheme of ^{89}Zr has been constructed up to spin $\frac{29}{2}^+$. The large-basis shell model code OXBASH was employed to analyze the level structure of ^{89}Zr. The results of shell-model calculation are in good accordance with the experiment.

Keywords: High-spin states; in-beam spectrum; shell model; level scheme.

1. Introduction

States of high momentum in nuclei could be constructed through intrinsic or collective modes of excitation. The former exists in the nuclei whose neutron or proton number near magic number, while the latter occurs in the nuclei whose neutron or proton number far away from magic number. As a special region in the nuclear structure studies, the $A \sim 90$ region has attracted lots of interests of both experimental and theoretical scientists. In this mass region for nuclei with N=48 to 50 there are four active orbitals, $1f_{5/2}$, $1g_{9/2}$, $2p_{1/2}$, $2p_{3/2}$, which may contribute to the state structure.[1] The level structure of ^{89}Zr was studied by different kinds of nuclear reaction. The low-lying states of ^{89}Zr was well studied using the reactions ^{91}Zr(p, t)^{89}Zr,^{86}Sr(α,nγ)^{89}Zr,^{83}Sr(α,2nγ)^{89}Zr[2,3], ^{88}Sr(α,3nγ)^{89}Zr[4]. In the

^{74}Ge(^{18}O,3nγ)^{89}Zr and ^{76}Ge(^{18}O,3nγ)^{89}Zr reactions, level scheme had been established.[5] The main purpose of the present work is to better characterize the high spin states of ^{89}Zr and to describe the level structure using shell model code OXBASH.

2. Experiment and results

The high-spin states of ^{89}Zr were populated via heavy ion fusion evaporation reaction ^{76}Ge(^{19}F,p5n)^{89}Zr at a bombarding energy of 80 MeV. The ^{19}F was delivered by the HI-13 tandem accelerator at the China Institute of Atomic Energy (CIAE). The target was prepared by vacuum evaporation of ^{76}Ge metal enriched to 96% to a thickness of 2.2 mg/cm^2 onto a 10 mg/cm^2 thick Pb backing.γ-γ coincidence events were obtained by using a multi-detector array consisting of 12 HPGe detectors with Compton suppression devices and 2 planar HPGe detectors to detect low energy γ rays. In the detector array, four detectors were placed at 90°, five at about 48° and five at 132° with respect to the beam direction. The energy resolution of each detectors was about 2 keV for 1332.5 keV γ rays. Gain matching and relative efficiency for these detectors were performed by standard ^{60}Co and ^{152}Eu γ sources. 68×10^6 γ-γ coincidence events were collected in the experiment in event-by-event mode.

γ-γ coincidence events were sorted offline into symmetrical matrices and directional correlation of oriented states (DCO) matrices. The DCO matrices were created by sorting the detectors lying at about 48°and 132° on one axis, while on the other axis were from the detectors placed at about 90° to the beam direction. These matrices were analyzed by Radware and Gaspware programs based on a Linux-PC computer. Relative intensities and DCO ratios of γ rays were calculated.

The level scheme of ^{89}Zr based on the present work is shown in Fig. 1(a). The majority of transitions observed in the study[5], have been confirmed. Three γ rays and two energy levels belonging to ^{89}Zr were identified in the work. The spins have been extended to $\frac{29}{2}^+$ with 6028.1 keV of excitation energy.

3. Shell model calculation

The nuclear shell model, which explains the regularities of the nuclear properties associated with the magic numbers, is one of the most successful model. Yrast band of ^{89}Zr was calculated by the shell model code OXBASH

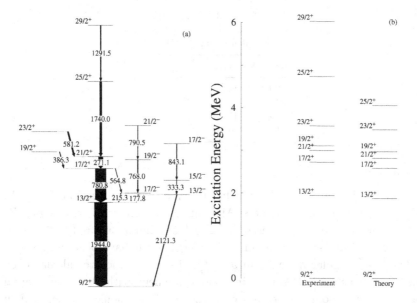

Fig. 1. Level scheme of ^{89}Zr induced in the present work (a) and comparison between theory and experiment (b).

based on shell model principles. The shell model Hamiltonian can be written as[1]

$$H = \sum_i \epsilon_i a_i^+ a_i + \sum_{ijkl} V_{ijkl} a_i^+ a_j^+ a_k a_l$$

For the two-body matrix elements V_{ijkl}, we use the residual interaction slgm[7]. ^{89}Zr has 40 protons and 49 neutrons. Therefore, ^{76}Sr was taken as inner core in the calculation. Occupation number was used to represent the particle number or hole number for a given single-particle orbital. The model space and occupation were given in Table. 1.

From the comparison between the shell model calculation and experimental results given in Fig. 1(b), we could conclude that the shell model calculation was in good accordance with the experimental results when angular momentum $J \leq \frac{23}{2}$. While $J \geq \frac{25}{2}$, there is a difference between the shell model calculation and the experimental results, by the reason that maybe some nucleons from other orbitals should be included in the calculation.

Table 1. Model space and nucleon occupation number used in the calculation.

Orbital	occupation number
$\pi 2p_{1/2}$	[0 2]
$\pi 1g_{9/2}$	[0 2]
$\nu 2p_{1/2}$	[1 2]
$\nu 1g_{9/2}$	[9 10]

4. Summary

The high-spin states of ^{89}Zr have been studied via the reaction ^{76}Ge(^{19}F, p5n)^{89}Zr at a beam energy of 80 MeV at CIAE. The level scheme of ^{89}Zr has been extended to $\frac{29}{2}^+$ based on the relative intensities and DCO ratios of γ rays. Three new γ rays and two energy levels have been assigned to ^{89}Zr. When angular momentum $J \leq \frac{23}{2}$, the level structure calculated by shell model code OXBASH is in good agreement with the experimental results. Maybe some nucleons from other orbitals should be included in the calculation, therefore the calculation did not well interpret the levels of angular momentum $J \geq \frac{25}{2}$. According to shell model calculation, ^{89}Zr is a spherical nucleus.

Acknowledgments

The authors are indebted to the staff of the HI-13 tandem accelerators at CIAE for providing good beam. The authors are also grateful to Professor G. J. Xu and Dr. Q. W. Fan for preparing the targets. This work is supported by National Natural Science Foundation of China (10175090, 10105015, 10375092) and by the Major State Basic Research Development Program under Grant No.2007CB815000.

References

1. P. Guazzoni et al., Nucl. Phys. A **697**, (2002) 611.
2. J. Bisping et al., Nucl. Phys. A **230**, (1974) 221.
3. T. Numao et al., J. Phys. Soc. Jpn. **47**, (1979) 365.
4. A. Nilsson M. Grecescu, Nucl. Phys. A **212**, (1973) 448.
5. E. K. Warburton et al., J. Phys. G: Nucl. Phys. **12**,(1986) 1017.
6. E. K. Warburton et al., Phys. Rev. C **31**, (1985) 1207.
7. F. J. D. Serduke, R. D. Lawon, Nucl. Phys. A **256**, (1976) 45.

CONSTRAINING THE SYMMETRY ENERGY FROM THE NEUTRON SKIN THICKNESS OF TIN ISOTOPES

LIE-WEN CHEN

Department of Physics, Shanghai Jiao Tong University, Shanghai 200240, China
E-mail: lwchen@sjtu.edu.cn

CHE MING KO*, JUN XU[†]

Cyclotron Institute and Department of Physics and Astronomy, Texas A&M University, College Station, Texas 77843-3366, USA
*E-mail: ko@comp.tamu.edu; †E-mail: xujun@comp.tamu.edu

BAO-AN LI

Department of Physics and Astronomy, Texas A&M University-Commerce, Commerce, Texas 75429-3011, USA
E-mail: Bao-An_Li@tamu-commerce.edu

We show in the Skyrme-Hartree-Fock approach that unambiguous correlations exist between observables of finite nuclei and nuclear matter properties. Using this correlation analysis to existing data on the neutron skin thickness of Sn isotopes, we find important constraints on the value $E_{\mathrm{sym}}(\rho_0)$ and density slope L of the nuclear symmetry energy at saturation density. Combining these constraints with those from recent analyses of isospin diffusion and double neutron/proton ratio in heavy ion collisions leads to a value of $L = 58 \pm 18$ MeV approximately independent of $E_{\mathrm{sym}}(\rho_0)$.

Keywords: Nuclear symmetry energy; neutron skin; Skyrme Hartree-Fock.

1. Introduction

The nuclear symmetry energy $E_{\mathrm{sym}}(\rho)$ plays a crucial role in both nuclear physics and astrophysics.[1,2] Although significant progress has been made in recent years in determining the density dependence of $E_{\mathrm{sym}}(\rho)$,[2] large uncertainties still exist even around the normal density ρ_0,[3] and this has hindered us from understanding more precisely many important properties of neutron stars.[4] To constrain the symmetry energy with higher accuracy is thus of crucial importance.

Theoretically, it has been established[5-13] that the neutron skin thickness $\Delta r_{np} = \langle r_n^2 \rangle^{1/2} - \langle r_p^2 \rangle^{1/2}$ of heavy nuclei, given by the difference of their neutron and proton root-mean-squared radii, provides a good probe of $E_{sym}(\rho)$. In particular, Δr_{np} has been found to correlate strongly with both $E_{sym}(\rho_0)$ and L in microscopic mean-field calculations.[5-11] It is, however, difficult to extract an accurate value for L from comparing the calculated Δr_{np} of heavy nuclei with experimental data as it depends on several nuclear interaction parameters in a highly correlated manner[7,8] and the calculations have been usually carried out by varying the interaction parameters. A well-known example is the Skyrme-Hartree-Fock (SHF) approach using normally 9 interaction parameters and there are more than 120 sets of Skyrme interaction parameters in the literature.

In the present talk, we report our recent work[14] on a new method to analyze the correlation between observables of finite nuclei and some macroscopic properties of asymmetric nuclear matter. Instead of varying directly the 9 interaction parameters within the SHF, we express them explicitly in terms of 9 macroscopic observables that are either experimentally well constrained or empirically well known. Then, by varying individually these macroscopic observables within their known ranges, we can examine more transparently the correlation of Δr_{np} with each individual observable. In particular, we have demonstrated that important constraints on $E_{sym}(\rho_0)$ and L can be obtained with the application of this correlation analysis to existing data on the neutron skin thickness of Sn isotopes.

2. The theoretical method

In the standard SHF model,[15] the 9 Skyrme interaction parameters, i.e., σ, $t_0 - t_3$, $x_0 - x_3$ can be expressed analytically in terms of 9 macroscopic quantities ρ_0, $E_0(\rho_0)$, the incompressibility K_0, the isoscalar effective mass $m_{s,0}^*$, the isovector effective mass $m_{v,0}^*$, $E_{sym}(\rho_0)$, L, the gradient coefficient G_S, and the symmetry-gradient coefficient G_V,[14,16] i.e.,

$$\sigma = \gamma - 1, \quad t_0 = 4\alpha/(3\rho_0), \quad x_0 = 3(y-1)E_{sym}^{loc}(\rho_0)/\alpha - 1/2,$$

$$t_1 = 20C/\left[9\rho_0(k_F^0)^2\right] + 8G_S/3, \quad x_1 = \left[12G_V - 4G_S - \frac{6D}{\rho_0(k_F^0)^2}\right]/(3t_1),$$

$$t_2 = \frac{4(25C - 18D)}{9\rho_0(k_F^0)^2} - \frac{8(G_S + 2G_V)}{3},$$

$$x_2 = \left[20G_V + 4G_S - \frac{5(16C - 18D)}{3\rho_0(k_F^0)^2}\right]/(3t_2),$$

$$t_3 = 16\beta/\left[\rho_0^\gamma(\gamma+1)\right], \quad x_3 = -3y(\gamma+1)E_{sym}^{loc}(\rho_0)/(2\beta) - 1/2, \quad (1)$$

where $k_F^0 = (1.5\pi^2\rho_0)^{1/3}$, $E_{\text{sym}}^{\text{loc}}(\rho_0) = E_{\text{sym}}(\rho_0) - E_{\text{sym}}^{\text{kin}}(\rho_0) - D$, and the parameters C, D, α, β, γ, and y are defined as[17]

$$C = \frac{m - m_{s,0}^*}{m_{s,0}^*} E_{\text{kin}}^0, \quad D = \frac{5}{9} E_{\text{kin}}^0 \left(4\frac{m}{m_{s,0}^*} - 3\frac{m}{m_{v,0}^*} - 1\right),$$

$$\alpha = -\frac{4}{3} E_{\text{kin}}^0 - \frac{10}{3} C - \frac{2}{3} [E_{\text{kin}}^0 - 3E_0(\rho_0) - 2C]$$
$$\times \frac{K_0 + 2E_{\text{kin}}^0 - 10C}{K_0 + 9E_0(\rho_0) - E_{\text{kin}}^0 - 4C},$$

$$\beta = \left[\frac{E_{\text{kin}}^0}{3} - E_0(\rho_0) - \frac{2}{3}C\right] \frac{K_0 - 9E_0(\rho_0) + 5E_{\text{kin}}^0 - 16C}{K_0 + 9E_0(\rho_0) - E_{\text{kin}}^0 - 4C},$$

$$\gamma = \frac{K_0 + 2E_{\text{kin}}^0 - 10C}{3E_{\text{kin}}^0 - 9E_0(\rho_0) - 6C}, \quad y = \frac{L - 3E_{\text{sym}}(\rho_0) + E_{\text{sym}}^{\text{kin}}(\rho_0) - 2D}{3(\gamma - 1)E_{\text{sym}}^{\text{loc}}(\rho_0)}, \quad (2)$$

with $E_{\text{kin}}^0 = \frac{3\hbar^2}{10m}\left(\frac{3\pi^2}{2}\right)^{2/3}\rho_0^{2/3}$ and $E_{\text{sym}}^{\text{kin}}(\rho_0) = \frac{\hbar^2}{6m}\left(\frac{3\pi^2}{2}\rho_0\right)^{2/3}$.

As a reference for the correlation analyses below, we use the MSL0 parameter set,[14] which is obtained by using the following empirical values for the macroscopic quantities: $\rho_0 = 0.16$ fm^{-3}, $E_0(\rho_0) = -16$ MeV, $K_0 = 230$ MeV, $m_{s,0}^* = 0.8m$, $m_{v,0}^* = 0.7m$, $E_{\text{sym}}(\rho_0) = 30$ MeV, and $L = 60$ MeV, $G_V = 5$ MeV·fm^5, and $G_S = 132$ MeV·fm^5. And the spin-orbital coupling constant $W_0 = 133.3$ MeV·fm^5 is used to fit the neutron $p_{1/2} - p_{3/2}$ splitting in ^{16}O. Using other Skyrme interactions obtained from fitting measured binding energies and charge rms radii of finite nuclei does not change our conclusion.

3. Results

To reveal clearly the dependence of Δr_{np} on each macroscopic quantity, we vary one quantity at a time while keeping all others at their default values in MSL0. Shown in Fig. 1 are the values of Δr_{np} for ^{208}Pb, ^{120}Sn and ^{48}Ca. Within the uncertain ranges for the macroscopic quantities considered here, the Δr_{np} of ^{208}Pb and ^{120}Sn exhibits a very strong correlation with L. However, it depends only moderately on $E_{\text{sym}}(\rho_0)$ and weakly on $m_{s,0}^*$. On the other hand, the Δr_{np} of ^{48}Ca displays a much weaker dependence on both L and $E_{\text{sym}}(\rho_0)$. Instead, it depends moderately on G_V and W_0. This explains the weaker Δr_{np}-$E_{\text{sym}}(\rho)$ correlation observed for ^{48}Ca in previous SHF calculations using different interaction parameters.[10]

Experimentally, the Δr_{np} of heavy Sn isotopes has been systematically measured.[18–23] As an illustration, we first show in the panel (a) of left

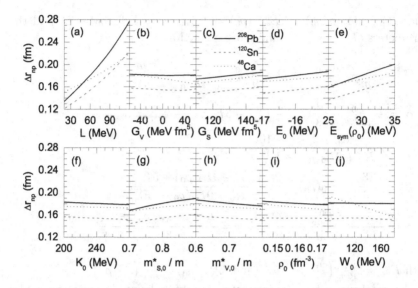

Fig. 1. (Color online) The neutron skin thickness Δr_{np} of ^{208}Pb, ^{120}Sn and ^{48}Ca from SHF with MSL0 by varying individually L (a), G_V (b), G_S (c), $E_0(\rho_0)$ (d), $E_{\rm sym}(\rho_0)$ (e), K_0 (f), $m_{s,0}^*$ (g), $m_{v,0}^*$ (h), ρ_0 (i), and W_0 (j). Taken from Ref.[14]

window in Fig. 2 the comparison of the available Sn Δr_{np} data with our calculated results using 20, 60 and 100 MeV, respectively, for the value of L and the default values for all other quantities in MSL0. It is seen that the value $L = 60$ MeV best describes the data. To be more precise, the χ^2 evaluated from the difference between the theoretical and experimental Δr_{np} values is shown as a function of L in the panel (b) of left window in Fig. 2. The most reliable value of L is found to be $L = 54 \pm 13$ MeV within a 2σ uncertainty.

Since the value of Δr_{np} depends on both L and $E_{\rm sym}(\rho_0)$, we have carried out a two-dimensional χ^2 analysis as shown by the grey band in the panel (c) of left window in Fig. 2. It is seen that increasing the value of $E_{\rm sym}(\rho_0)$ systematically leads to smaller values of L. Furthermore, we have estimated the effects of nucleon effective mass by using $m_{s,0}^* = 0.7m$ and $m_{v,0}^* = 0.6m$ as well as $m_{s,0}^* = 0.9m$ and $m_{v,0}^* = 0.8m$, in accord with the empirical constraint $m_{s,0}^* > m_{v,0}^*$,[2,3,24] and the resulting constraints are shown by the dashed and dotted lines. As expected from the results shown in Fig. 1, effects of nucleon effective mass are small with the value of L shifting by only a few MeV for a given $E_{\rm sym}(\rho_0)$. As one has also expected, effects of varying other macroscopic quantities are even smaller.

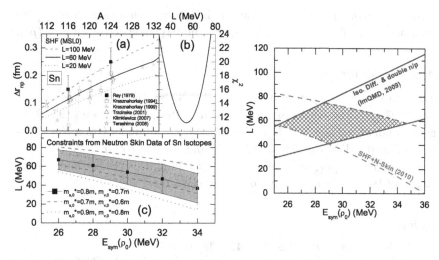

Fig. 2. (Color online) Left window: (a) The Δr_{np} data for Sn isotopes from different experimental methods and results from SHF calculation using MSL0 with $L = 20$, 60 and 100 MeV. (b) χ^2 as a function of L. (c) Constraints on L and $E_{\mathrm{sym}}(\rho_0)$ from the χ^2 analysis of the Δr_{np} data on Sn isotopes (Grey band as well as dashed and dotted lines). Right window: Constraints on L and $E_{\mathrm{sym}}(\rho_0)$ obtained in the present work (dashed lines) and that from Ref.[25] (solid lines). The shaded region represents their overlap. Taken from Ref.[14]

The above constraints on the L-$E_{\mathrm{sym}}(\rho_0)$ correlation can be combined with those from recent analyses of isospin diffusion and double n/p ratio in heavy ion collisions at intermediate energies[25] to determine simultaneously the values of both L and $E_{\mathrm{sym}}(\rho_0)$. Shown in the right window of Fig. 2 are the two constraints in the $E_{\mathrm{sym}}(\rho_0)$-L plane. Interestingly, these two constraints display opposite L-$E_{\mathrm{sym}}(\rho_0)$ correlations. This allows us to extract a value of $L = 58 \pm 18$ MeV approximately independent of the value of $E_{\mathrm{sym}}(\rho_0)$. This value of L is quite precise compared to existing estimates in the literature (See Ref.[3] for a recent summary) although the constraint on $E_{\mathrm{sym}}(\rho_0)$ is not improved.

4. Summary

We have proposed a new method to analyze the correlations between observables of finite nuclei and some macroscopic properties of nuclear matter, and demonstrated that the existing neutron skin data on Sn isotopes can give important constraints on the symmetry energy parameters L and $E_{\mathrm{sym}}(\rho_0)$. Combining these constraints with those from recent analyses of isospin diffusion and double n/p ratio in heavy ion collisions leads to a quite accurate value of $L = 58 \pm 18$ MeV approximately independent of $E_{sym}(\rho_0)$.

Acknowledgments

This work was supported in part by the NNSF of China under Grant No. 10975097, the National Basic Research Program of China (973 Program) under Contract No. 2007CB815004, U.S. NSF under Grant No. PHY-0758115 and PHY-0757839, the Welch Foundation under Grant No. A-1358, the Research Corporation under Award No. 7123, the Texas Coordinating Board of Higher Education Award No. 003565-0004-2007.

References

1. A. W. Steiner *et al.*, *Phys. Rep.* **411**, 325 (2005).
2. B. A. Li, L. W. Chen, and C. M. Ko, *Phys. Rep.* **464**, 113 (2008).
3. C. Xu, B. A. Li, and L. W. Chen, *Phys. Rev.* **C82**, 054607 (2010).
4. J. Xu *et al.*, *Phys. Rev.* **C79**, 035802 (2009); *Astrophys. J.* **697**, 1549 (2009).
5. B. A. Brown, *Phys. Rev. Lett.* **85**, 5296 (2000); S. Typel and B. A. Brown, *Phys. Rev.* **C64**, 027302 (2001).
6. C. J. Horowitz and J. Piekarewicz, *Phys. Rev. Lett.* **86**, 5647 (2001).
7. R. J. Furnstahl, *Nucl. Phys.* **A706**, 85 (2002).
8. P.–G. Reinhard and W. Nazarewicz, *Phys. Rev.* **C81**, 051303 (2010).
9. S. Yoshida and H. Sagawa, *Phys. Rev.* **C69**, 024318 (2004); *ibid.* **C73**, 044320 (2006).
10. L. W. Chen, C. M. Ko, and B. A. Li, *Phys. Rev.* **C72**, 064309(2005).
11. B. G. Todd-Rutel and J. Piekarewicz, *Phys. Rev. Lett.* **95**, 122501 (2005).
12. P. Danielewicz, *Nucl. Phys.* **A727**, 233 (2003).
13. M. Centelles *et al.*, *Phys. Rev. Lett.* **102**, 122502 (2009); M. Warda *et al.*, *Phys. Rev.* **C80**, 024316 (2009).
14. L. W. Chen *et al.*, *Phys. Rev.* **C82**, 024321 (2010).
15. E. Chabanat *et al.*, *Nucl. Phys.* **A627**, 710 (1997).
16. L. W. Chen, arXiv:1101.2384.
17. L. W. Chen *et al.*, *Phys. Rev.* **C80**, 014322 (2009); L. W. Chen, *Sci. China Ser.* **G52**, 1494 (2009) [arXiv:0911.1092].
18. L. Ray *et al.*, *Phys. Rev.* **C19**, 1855 (1979).
19. A. Krasznahorkay *et al.*, *Nucl. Phys.* **A567**, 521 (1994).
20. A. Krasznahorkay *et al.*, *Phys. Rev. Lett.* **82**, 3216 (1999).
21. A. Trzcinska *et al.*, *Phys. Rev. Lett.* **87**, 82501 (2001).
22. A. Klimkiewicz *et al.*, *Phys. Rev.* **C76**, 051603(R) (2007).
23. S. Terashima *et al.*, *Phys. Rev.* **C77**, 024317 (2008).
24. T. Lesinski *et al.*, *Phys. Rev.* **C74**, 044315 (2006).
25. M. B. Tsang *et al.*, *Phys. Rev. Lett.* **102**, 122701 (2009).

WOBBLING ROTATION IN ATOMIC NUCLEI*

Y. S. CHEN[1,2] and ZAO-CHU GAO[1]

[1]*China Institute of Atomic Energy, P.O. Box 275(18) Beijing 102413, China*
[2]*Institute of Modern Physics, Chinese Academy of sciences Lanzhou 730000, China*
E-mail: yschen@ciae.ac.cn

The wobbling states are described microscopically with the Triaxial Projected Shell Model (TPSM), and the experimental wobbling bands firmly established in [163]Lu have been well reproduced by the present calculation. By using calculated wave functions of wobbling states, the projections of the total angular momentum along the three body-fixed axes are calculated as functions of spin. This calculation results in the dynamic geometry of the angular momentum in the intrinsic frame, and thus promises to provide insight into the wobbling motion in nuclei. The similar TPSM calculations have been performed to study the possible wobbling excitation in the neighboring nucleus [164]Lu. It has been found that the wobbling structure in odd-odd Lu may be modified by the extra neutron, but not completely destroyed, indicating the possibility for observing the wobbling bands in the odd-odd Lu isotopes.

1. Introduction

More general schemes of rotational motion suggested by classical mechanics are precession and wobbling,[1] henceforth both are called wobbling. The wobbling motion in nuclei is the excitation mode associated with the rotational asymmetry of a triaxial quantum system, which was proposed first by Bohr and Mottelson in the early 1970s.[2,3] It is expected that in a rotating stable triaxial nucleus the angular momentum is not aligned with any of the body-fixed axes and rather processes and wobbles around one of these axes in a manner just like that of an asymmetric top. Although the wobbling rotation is expected to be a general excitation mode that would occur in triaxially deformed nuclei, it had never been realized in experimental nuclear spectra until the clear evidence was found for [163]Lu in 2001. The nuclear

*This work is supported by the National Natural Science Foundation of China under Grant No.s 10775182, 11021504 and by the Chinese Major State Basic Research Development Program through Grant No. 2007CB815000.

wobbling mode has been firmly established by the crucial experiments that discovered the first- and second-phonon wobbling bands in ^{163}Lu.[4-6] The subsequent experiments for searching other wobblers have been performed quickly over the past decade and the wobbling bands have been seen in only a few Lu isotopes, namely, in ^{161}Lu,[7] ^{165}Lu[8,9] and ^{167}Lu.[10] In spite of many experiments, e.g., in Refs.[11-13] searching for wobbling modes in nearby nuclei there are no more further examples of wobbling bands. Only very recently, the wobbling bands have been observed for the first time in a nucleus other than Lu, namely, in ^{167}Ta, making this collective triaxial rotational mode a more general phenomenon.[14]

It is an essential point in the study of nuclear rotations to establish whether and how the collective wobbling motion is generated microscopically by the nucleons. One of best approaches to provide insight into this quantum many-body problem is the shell model. It is likely that the wobbler ^{163}Lu may serve as the best target nucleus for such a theoretical study since its rotational spectra have been measured up to the second phonon band and the configurations of the wobbling bands well assigned, which provide a complete information about the phonon energies and the intrinsic structure. In the present paper, the wobbling states are described microscopically by the Triaxial Projected Shell Model(TPSM). We will see that the experimental wobbling bands observed in ^{163}Lu can be well reproduced by the present calculation. We also show that a deep understanding of the wobbling motion in nuclei can be achieved by analyzing the dynamic geometry of the angular momentum in the intrinsic frame, which can be calculated by using the TPSM wave functions. To investigate the possible wobbling excitation in odd-odd Lu nuclei the similar TPSM calculation has been performed for the neighboring nucleus ^{164}Lu, in which an extra neutron is added to a wobbler ^{163}Lu. It has been found that the wobbling structure may be modified by the addition of the extra neutron in some degrees, but not completely destroyed , indicating the possibility to observe the wobbling bands in the odd-odd Lu isotopes.

The model used is briefly described in Section 2. The TPSM calculations for the wobbling bands and discussions are given in Section 3. General conclusions are summarized in Section 4.

2. Brief description of the model

The theory incorporates the many-body correlations into the mean-field approximations to make shell-model calculations possible for heavy nuclei. The present model employs a triaxially deformed basis and constructs the model space by including multi-quasi-particle (qp) states and performs exact three-dimensional angular momentum projection. A realistic two-body Hamiltonian is then diagonalized in this space. The theory is briefly described below and the more details can be found in Ref.[16] In the TPSM, the trial wave function may be written as

$$|\Psi_{IM}^{\sigma}\rangle = \sum_{K\kappa} f_{IK_\kappa}^{\sigma} \hat{P}_{MK}^{I} |\Phi_\kappa\rangle, \qquad (1)$$

where $\hat{P}_{MK}^{I}$ is the three-dimensional angular-momentum-projection operator,

$$\hat{P}_{MK}^{I} = \frac{2I+1}{8\pi^2} \int d\Omega\, D_{MK}^{I}(\Omega)\, \hat{R}(\Omega), \qquad (2)$$

where $R(\Omega)$ is the rotation operator which has the explicit form, $e^{-\imath\alpha\hat{J}_z}e^{-\imath\beta\hat{J}_y}e^{-\imath\gamma\hat{J}_z}$, Ω represents a set of Euler angles ($\alpha, \gamma = [0, 2\pi]$ and $\beta = [0, \pi]$). The σ in Eq. (1) specifies the states with the same angular momentum I. The dimension of the summation in Eq. (1) is $K \times \kappa$, where $|K| \leq I$ and κ is usually in the order of 10^2. The $|\Phi_\kappa\rangle$ represents a set of multi-qp states associated with the triaxially deformed qp vacuum $|0\rangle$. For odd proton nuclei included is the set of 1-, 3-, and 5-qp states,

$$\{\alpha_{\pi_1}^{\dagger} |0\rangle,\quad \alpha_{\nu_1}^{\dagger}\alpha_{\nu_2}^{\dagger}\alpha_{\pi_1}^{\dagger} |0\rangle,\quad \alpha_{\nu_1}^{\dagger}\alpha_{\nu_2}^{\dagger}\alpha_{\pi_1}^{\dagger}\alpha_{\pi_2}^{\dagger}\alpha_{\pi_3}^{\dagger} |0\rangle\}. \qquad (3)$$

By carrying out the variational procedure with respect to the wave function, precisely the coefficients $f_{IK_\kappa}^{\sigma}$, we obtain the eigenvalue equation,

$$\sum_{K\kappa} f_{IK_\kappa}^{\sigma} (\langle\Phi_\kappa|HP_{K'K}^{I}|\Phi_\kappa\rangle - E\langle\Phi_\kappa|P_{K'K}^{I}|\Phi_\kappa\rangle) = 0. \qquad (4)$$

The shell model Hamiltonian considered involves a large number of nucleons moving in a spherical Nilsson potential and an interaction of separable multipole $Q \cdot Q$ plus monopole pairing plus quadrupole pairing,

$$H = H_0 - \frac{1}{2}\sum_{\lambda=2}^{4} \chi_\lambda \sum_{\mu=-\lambda}^{\lambda} Q_{\lambda\mu}^{\dagger}Q_{\lambda\mu}$$

$$- G_0 P_{00}^{\dagger}P_{00} - G_2 \sum_{\mu=-2}^{2} P_{2\mu}^{\dagger}P_{2\mu}. \qquad (5)$$

In Eq. (5), $\hat{H}_0$ is the spherical single-particle Hamiltonian, which contains a proper spin-orbit force.[17] The second term is quadrupole-quadrupole (QQ) interaction that includes the nn, pp, and np components. The QQ interaction strength χ is determined in a self-consistent way to match the quadrupole deformation used in construction of the qp basis. The third term in Eq. (5) is monopole pairing, whose strength G_M is of the standard form G/A, with $G = 17.93$ MeV for both protons and neutrons , which approximately reproduces the observed odd-even mass differences in this mass region. The last term is quadrupole pairing, with the strength $G_Q=0.12G_M$ as usual.

The triaxially deformed single particle states are generated by the Nilsson Hamiltonian

$$\hat{H}_N = \hat{H}_0 - \frac{2}{3}\hbar\omega\epsilon_2\left(\cos\gamma\hat{Q}_0 - \sin\gamma\frac{\hat{Q}_{+2} + \hat{Q}_{-2}}{\sqrt{2}}\right), \qquad (6)$$

where the parameters ϵ_2 and γ describe quadrupole deformation and triaxial deformation, respectively.

3. Results of calculations and discussion

The three major shells of $N = 4, 5, 6$ for both neutrons and protons are included to calculate the Nilsson single particle states. The deformation parameters used to generate the deformed quasiparticle basis states are $\varepsilon_2=0.38$, $\gamma = 20°$ and $\varepsilon_4=0.0$ for ^{163}Lu, which are in consistent with the TRS calculation. The shell model basis contains the multi-quasiparticle configurations up to the 5-qp states of 2n3p, which has a seniority quantum number high enough to describe the rotational bands up to very high spin $I \sim 101/2$, the highest spin observed in ^{163}Lu. The calculated wobbling bands of ^{163}Lu up to second phonon states are shown in Figure 1, and compared with the experimental data.[6] In Figure 1, the excitation energy minus a rigid-rotor reference is plotted as a function of spin for the triaxial strongly formed bands, TSD1, TSD2 and TSD3, which are identified as zero phonon, one phonon and two phonon wobbling bands, respectively. The collective wobbling behavior can be described within a phonon model, where the energy of each band is expressed as $E = \frac{\hbar^2}{2J}I(I+1) + \hbar\omega_{\text{wob}}(n_w + 1/2)$, where the n_w is phonon quantum number and ω_{wob} is the wobbling frequency that measures the wobbling phonon excitation energy. It is seen from figure 1 that the experimental energies of the all three bands of the wobbling family are well reproduced by the present calculations. The experimental excitation energies of the one phonon band and the two phonon band relative to the zero phonon band, and thus the wobbling frequencies ω_{wob}, have been also well reproduced by these calculations. It is shown that the observed excitation energy of the TSD3 band is not twice as large as the excitation energy of the TSD2 band in the range of observed angular

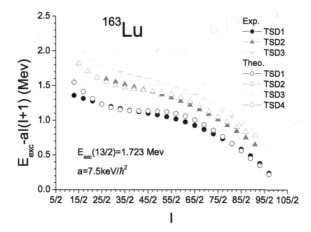

Fig. 1. Excitation energy minus a rigid-rotor reference for denoted TSD bands in ^{163}Lu.
The inertia parameter was set to 7.5 keV/$\hbar^2$ and the TSD1 band head at $I = 13/2$ was
set to E_{exc}=1.723 MeV.The calculated results (open symbols) are compared with the
experimental data (solid symbols).

momentum, indicating the anharmonicity of the wobbling excitation. The
presence of the anharmonicity implies the existence of the complex coupling between the collective wobbling motion and the single quasiparticle
excitations in a rapidly rotating triaxial nucleus. Nevertheless, the effect of
anharmonicity on the wobbling excitation energy has been well described
by the TPSM theory. The anharmonicity of the wobbling motion in the $E2$
transitions was studied in detail by using the particle-rotor model of one
$i_{13/2}$ quasiparticle coupled to the core of triaxial shape.[18]

It is an advantage of the TPSM theory that the wave function in the
form of angular momentum projection can be used conveniently to calculate the dynamic geometry of the total angular momentum in the intrinsic
frame. Such a calculation is particularly useful in understanding the wobbling motion in nuclei because the family of wobbling bands is characterized
with the common similar intrinsic structure but the different tilting of the
angular momentum vector with respect to one of body-fixed axes. In Figure
2, presented are the expectation values of I_x^2, I_y^2 and I_z^2, divided by $I(I+1)$,
as functions of spin for the lowest two TSD bands in ^{163}Lu, calculated by
using the TPSM wave functions which reproduce the experimental band energies as shown in Figure 1. The TSD1 and TSD2 are the signature partner
bands and their spin sequences can be expressed by means of the signa-
ture α as $I = 2n + \alpha$ with the favored signature $\alpha=1/2$ and the unfavored
signature $\alpha=-1/2$, respectively. It is seen from Figure 2 that the favored

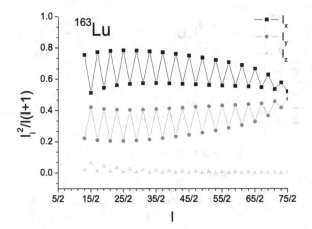

Fig. 2. Expectation values of I_x^2, I_y^2 and I_z^2, divided by $I(I + 1)$, as functions of spin for the lowest two TSD bands in ^{163}Lu, calculated by using the TPSM wave functions that reproduce the experimental band energies as shown in Fig. 1.

signature band TSD1, I=13/2, 17/2, 21/2, ..., has the angular momentum vector most closing to the x-axis, the shortest body-fixed axis, and thus the rotational band is energetically lowest and assigned as the n_ω=0 phonon state. While in the unfavored signature band TSD2, I=15/2, 19/2, 23/2, ..., the angular momentum vector lies farther from the x-axis and thus the band is relatively higher in energy and assigned as the n_ω=1 phonon state. The calculated wave functions show a very similar structure for the wobbling bands, namely, the $i_{13/2}$ proton qp-configuration dominates in a large range of spin. The very similar intrinsic structure together with the tilting motion of the angular momentum vector, as being demonstrated in Figure 1, characterize the wobbling mode in nuclei. The stable triaxial deformation and the high-j and low-K orbit, like $\pi[660]1/2$ in ^{163}Lu, play an essential role in the formation of the wobbling bands. For the wobbling excitation the crucial contribution from the aligned particle was also born out from the calculation of the wobbling mode in ^{163}Lu by means of the cranking shell model plus random phase approximation.[19]

To date no any experimental evidence for the wobbling mode has been found in odd-odd nuclei. To investigate this possibility the similar TPSM calculation has been performed for nucleus ^{164}Lu, in which an extra neutron is added to the wobbler ^{163}Lu. The deformation parameters used to generate the deformed quasi-particle basis states are ε_2=0.41, $\gamma = 21°$ and ε_4=0.0, which are in consistent with the TRS calculation. With these deformations the calculated yrast TSD band is in agreement with the experimental TSD1 band in ^{164}Lu. In order to make the result more instructive the calculation of the geometry of the angular momentum has been performed for the single 2qp configuration of $\pi[660]1/2\nu[523]5/2$, a proton in the $i_{13/2}$ and a neutron in the $f_{7/2}$ shells, which dominates the TPSM wave functions in a large range of spin. Calculated expectation values of I_x^2, I_y^2 and I_z^2,divided by $I(I+1)$, as functions of spin for the lowest two TSD bands in ^{164}Lu are shown in Figure 3. The characteristic wobbling motion remains in ^{164}Lu

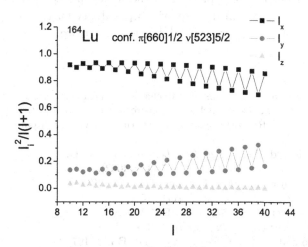

Fig. 3.   Calculated expectation values of I_x^2, I_y^2 and I_z^2, divided by $I(I+1)$, as functions of spin for the lowest TSD bands in ^{164}Lu by using the TPSM wave functions for the single 2qp configuration of $\pi[660]1/2\nu[523]5/2$.

as it is shown in Figure 3 that in going from the $n_\omega=0$ band, $I=10$, 12, 14, ..., to the $n_\omega=1$ band, $I=11$, 13, 15, ..., the angular momentum vector lies progressively farther from the x-axis, just as do in the case of ^{163}Lu. It seems to be reasonable to expect that the wobbling mode presented in odd-A Lu nuclei may be modified by the addition of the extra neutron but not completely destroyed , indicating the possibility for the observation of the wobbling bands in the odd-odd Lu isotopes.

4. Summary

The wobbling rotation in atomic nuclei can be described full quantum-mechanically by mean of the TPSM theory. The wobbling bands observed in ^{163}Lu have been well reproduced by the present calculation up to very high spins and the second wobbling phonon excitations within the shell model space that spans the set of 1qp, 3qp and 5qp states. The wobbling motion has been vividly demonstrated with the dynamic geometry of the angular momentum in the body-fixed frame, which may be calculated by using the TPSM wave functions. The wobbling excitation is characterized with the common similar intrinsic structure, particularly the stable triaxial shape, and the tilting motion of the angular momentum vector in the intrinsic frame where the vector lies progressively farther from the shortest x-axis when going from the zero phonon to the higher phonon numbers. Once these conditions are fulfilled there exist a possibility for the observation of new wobbling nuclei, for example, ^{164}Lu is a candidate.

References

1. L. D. Landau and E. M. Lifshitz, Mechanics, p117, Pergamon, London (1960).
2. B. R. Mottelson, Proc. Nuclear Structure Symp., Joutsa (1970).
3. A. Bohr and B. R. mottelson , Nuclear Structure, Vol.2, Benjamin, New York (1975).
4. S. W. Ødegård et al., Phys. Rev. Lett. **86** 5866(2001).
5. D. R. Jensen et al., Nucl. Phys. **A703**, 3(2002).
6. D. R. Jensen et al., Phys. Rev. Lett. **89**, 142503(2002).
7. P. Bringel et al., Eur. Phys. J. **A16**, 155(2003).
8. G. Schönwaßer et al., Phys. Lett. **B552**, 9(2003).
9. G. Schönwaßer et al., Nucl. Phys. **A735**,393(2004).
10. H. Amro and W. C. Ma et al.,Phys. Lett. **B553**, 197(2003).
11. Y. C. Zhang and W. C. Ma et al., Phys. Rev. **C76**, 064321(2007).
12. D. J. Hartley and M. K. Djongolov et al., Phys. Lett. **B 608**, 31(2005).
13. D. T. Scholes and D. M. Cullen et al., Phys. Rev. **C70**, 054314(2004).
14. D. J. Hartley and R. V. Janssens et al., Phys. Rev. **C80**, 041304(R)(2009).
15. B. Crowell et al., Phys. Rev. Lett. **72**, 1164(1994).
16. Zho-Chu Gao, Y. S. Chen and Yang Sun, Phys. Lett. **B634**, 195(2006).
17. T. Bengtsson and I. Ragnarsson, Nucl. Phys. **A436**, 14(1985).
18. I. hamamoto, Phys. Rev. **C65**, 044305(2002).
19. M. Matsuzaki, Y. R. Shimizu and K. Matsuyanagi, Phys. Rev. **C65**, 041303(R)(2002).

THE MIXING OF SCALAR MESONS AND THE POSSIBLE NONSTRANGE DIBARYONS

L. R. DAI,* X. S. KANG, S. X. AO, Y. WANG, C. B. WANG and W. W. SUN

Department of Physics, Liaoning Normal University, Dalian, 116029, China
*E-mail: dailianrong@gmail.com

The mixing of scalar mesons is an open problem. In this work, the idea of the mixing of scalar mesons will be applied to dibaryon system on quark level. By introducing the mixing of scalar mesons, we dynamically investigate the structure of possible nonstrange dibaryons in the chiral $SU(3)$ quark model by solving the resonating group method equation. The results show that no matter what kind of mixing is considered, the binding energies of nonstrange dibaryons would become stable if reasonable parameters are used.

Keywords: Quark model; chiral symmetry; dibaryon; the mixing of scalar mesons.

1. Introduction

The mixing of scalar mesons is an open problem, since the structure of scalar meson is unclear and controversial.[1] Now we will apply the idea of the mixing of scalar mesons to dibaryon system.

There are many works to study the dibaryon system.[2–11] In this work, we will focus the nonstrange dibaryon system on quark level. In 1987, Yazaki analyzed systems with two nonstrange baryons in the framework of the cluster model. The author considered the one-gluon exchange (OGE) interaction and the confining potential between two quarks and showed that $\Delta\Delta$ system could be bound state because the color magnetic interaction between two clusters is attractive.[2] In the quark delocalization model, the authors studied the structure of the $\Delta\Delta$ system and found that this system is a deeply bound state.[3–5] In the chiral $SU(3)$ quark model,[6] the authors also studied this system and found this system is also a bound state,[8] but the binding energy is not as large as those in the quark delocalization model. However, the mixing of scalar mesons is never considered for dibaryon systems in above works. Therefore, in the present work, we would like to further study these nonstrange dibaryon systems.

Among the quark models, one of the most successful models is the chiral $SU(3)$ quark model.[6] In this model, the source of the constituent quark mass can be logically explained from the underlying chromodynamics (QCD) theory of the strong interaction. Since spontaneous vacuum breaking has to be considered, and as a consequence the coupling between the quark field and the Goldstone boson must be introduced to restore the chiral symmetry. In this sense, the chiral quark model can be regarded as a quite reasonable and useful model to describe the medium-range nonperturbative QCD effect. This model has been quite successful in reproducing the energies of the baryon ground states, the binding energy of deuteron, the nucleon-nucleon (NN) scattering phase shifts of different partial waves, and the hyperon-nucleon (YN) cross sections by solving the resonating group method (RGM) equation.

Recently, using this chiral $SU(3)$ quark model, the author firstly introduced the mixing of scalar mesons into NN system[12] and YN system.[13] In Ref.[12] showed that no matter what kind of mixing is taken, the scattering phase shifts of nucleon-nucleon system can be reasonably reproduced in the chiral $SU(3)$ quark model. Inspired by this, by introducing the mixing of scalar mesons and using the same set of parameters used in NN scattering processes,[12] we would like to further investigate the structure of possible nonstrange dibaryons.

2. Formulation

The chiral $SU(3)$ quark model has been described in the literature[6] and we refer the reader to the work for details. Here we give the salient feature of this model. In the chiral $SU(3)$ quark model, the coupling between chiral field and quark is introduced to describe nonperturbative QCD effect. The interacting Lagrangian can be written as:

$$\mathcal{L}_I = -g_{ch}\overline{\psi}(\sum_{a=0}^{8} \sigma^a \lambda_a + i \sum_{a=0}^{8} \pi_a \lambda_a \gamma_5)\psi, \tag{1}$$

where λ_0 is a unitary matrix, $\sigma_0, \ldots, \sigma_8$ are the scalar nonet field, and $\pi_0, \ldots, \pi_8$ the pseudoscalar nonet fields. The $\mathcal{L}_I$ is invariant under the infinitesimal chiral $SU(3)_L \times SU(3)_R$ transformation, and only one coupling constant g_{ch} is needed by chiral symmetry requirement.

The total hamiltonian of baryon-baryon systems can be written as

$$H = \sum_{i=1}^{6} T_i - T_G + \sum_{i<j=1}^{6} V_{ij}, \tag{2}$$

$$V_{ij} = V_{ij}^{conf} + V_{ij}^{OGE} + V_{ij}^{ch}, \tag{3}$$

where $\sum_i T_i - T_G$ is the kinetic energy of the system, and V_{ij} includes all interactions between two quarks. V_{ij}^{conf} is the confinement potential taken as quadratic form, V_{ij}^{OGE} is the OGE interaction, and V_{ij}^{ch} represents the chiral fields induced effective quark-quark potential, which includes the scalar boson exchanges and the pseudoscalar boson exchange,

$$V_{ij}^{ch} = \sum_{a=0}^{8} V_{ij}^{\sigma_a} + \sum_{a=0}^{8} V_{ij}^{\pi_a}. \tag{4}$$

The detailed formula expressions can be found in Ref.[6]

The definition of mixing of scalar mesons is:

$$\sigma = \sigma_8 \sin\theta_s + \sigma_0 \cos\theta_s,$$
$$\epsilon = \sigma_8 \cos\theta_s - \sigma_0 \sin\theta_s. \tag{5}$$

here the mixing angle of scalar singlet and octet mesons is θ_s. Three cases will be discussed: the first is for no mixing, where the mixing angle is zero; the second is general mixing where the mixing angle is $\theta_s = -18°$ introduced in Ref.,[14] based on investigation of a dynamically spontaneous symmetry breaking mechanism; and the third is ideal mixing where the mixing angle is 35.3°, which means that σ only acts on the $u(d)$ quark, and ϵ meson on the s quark.

Now we briefly give the procedure for the parameters determination. The initial input parameters are taken to be the usual values: i.e, the harmonic oscillator width parameter $b_u = 0.5$ fm, the up (down) quark mass $m_{u(d)} = 313$ MeV, the coupling constant for scalar and pseudoscalar chiral field coupling, g_{ch}, is fixed by the relation

$$\frac{g_{ch}^2}{4\pi} = \frac{9}{25} \frac{m_u^2}{M_N^2} \frac{g_{NN\pi}^2}{4\pi} \tag{6}$$

with the experimental value $g_{NN\pi}^2/4\pi = 13.67$. The mass of the mesons are taken to be experimental values, except for the σ meson. The cutoff mass is taken to be the value close to the chiral symmetry breaking scale.[15–18] The OGE coupling constants g_u and the strengths of the confinement potential a_{uu} and a_{uu}^0 are determined by baryon masses and their stability conditions. The model parameters are listed in Table 1.

3. Resonating group method (RGM)

Resonating group method is a well established method for studying the interaction between two clusters.

The total, antisymmetrized six-quark wave function in orbital, spin, flavor and color space is of the following form:

$$\Psi = \mathcal{A}[\phi_A(\boldsymbol{\xi}_1, \boldsymbol{\xi}_2)\phi_B(\boldsymbol{\xi}_3, \boldsymbol{\xi}_4)\chi(\boldsymbol{R}_{AB})Z(\boldsymbol{R}_{CM})]_{STC}, \tag{7}$$

Table 1. Model parameters. Meson masses and cutoff masses: m_π = 138 MeV, m_K = 495 MeV, m_η = 549 MeV, $m_{\eta'}$ = 957 MeV, $m_{\sigma'}$ = m_ϵ = m_κ = 980 MeV, Λ = 1100 MeV for all mesons.

	Set I	Set II	Set III
θ^s	0°	−18°	35.3°
b_u(fm)	0.5	0.5	0.5
m_u(MeV)	313	313	313
g_{ch}	2.621	2.621	2.621
m_σ (MeV)	594	491	559
g_u^2	0.785	0.785	0.785
a_{uu}(MeV/fm^2)	48.1	49.0	50.0
a_{uu}^0(MeV/fm^2)	-43.6	-44.9	-46.0

where ξ_1, ξ_2 are the internal coordinates for the cluster A, and ξ_3, ξ_4 are the internal coordinates for the cluster B. R_{AB} is the relative coordinate between A and B, and R_{CM} is the center of mass coordinate of the total system. S and T denote the total spin and isospin of the cluster A and cluster B, and the ϕ_A and ϕ_B are the antisymmetrized wave functions of cluster A and B, respectively. The $\chi(R_{AB})$ is the relative wave function of the two clusters, and $Z(R_{CM})$ is the total center of mass wave function of the six-quark state which can be choosen freely due to the Galilei invariance of the system. The symbol $\mathcal{A}$ is the antisymmetrizing operator defined as

$$\mathcal{A} \equiv 1 - \sum_{i\in A, i\in B} P_{ij}. \tag{8}$$

where P_{ij} is the permutation operator of the ith and jth quarks.

Substituting Ψ into the projection equation

$$\langle \delta\Psi|(H - E)|\Psi\rangle = 0, \tag{9}$$

we obtain the coupled integro-differential equation for the relative function χ as

$$\int [\mathcal{H}(R, R') - E\mathcal{N}(R, R')]\chi(R')dR' = 0, \tag{10}$$

where the Hamiltonian kernel $\mathcal{H}$ and normalization kernel $\mathcal{N}$ can, respectively, be calculated by

$$\left\{ \begin{matrix} \mathcal{H}(R, R') \\ \mathcal{N}(R, R') \end{matrix} \right\} = \left\langle [\hat{\phi}_A(\xi_1, \xi_2)\hat{\phi}_B(\xi_3, \xi_4)]\delta(R - R_{AB}) \right.$$

$$\left| \left\{ \begin{matrix} H \\ 1 \end{matrix} \right\} \right| \left. \mathcal{A} \left[[\hat{\phi}_A(\xi_1, \xi_2)\hat{\phi}_B(\xi_3, \xi_4)]\delta(R' - R_{AB}) \right] \right\rangle.$$

$$\tag{11}$$

Equation (11) is the so-called coupled-channel RGM equation. Expanding unknown $\chi(\boldsymbol{R}_{AB})$ by employing well-defined basis wave functions, such as Gaussian functions, one can solve the coupled-channel RGM equation for a bound-state problem to obtain the binding energy for the two-cluster systems. The details of solving the RGM equation can be found in Refs.[19–23]

4. Result

Now three different states will be selected. One state is NN system with $L = 0, S = 1, T = 0$, this is the only dibaryon state confirmed by experiment, we call it deuteron. Other two states are from $\Delta\Delta$ system: one is the state with $L = 0, S = 0, T = 3$, in which only the central force is needed; and another state is called deltaron, $L = 0, S = 3, T = 0$, in which the hidden color channel (CC) is included.

$$|CC\rangle = -\frac{1}{2}|\Delta\Delta\rangle + \frac{\sqrt{5}}{2}\mathcal{A}_{STC}|\Delta\Delta\rangle \qquad (12)$$

where $\mathcal{A}_{STC}$ stands for the antisymmetrizer in the spin-isospin-color space.

Using the model parameters listed in Table 1, we dynamically investigate the structure of possible nonstrange dibaryons in the chiral $SU(3)$ quark model by solving the RGM equation. These model parameters can give a good description of the energies of the baryon ground states, the binding energy of deuteron, and the experimental data of the nucleon-nucleon scattering processes. The calculated

Table 2. The calculated binding energy for three different dibaryons (MeV).

	$NN (ST = 10)$	$\Delta\Delta (ST = 03)$	$\Delta\Delta (ST = 30)$
Set I	2.21	22.46	48.53
Set II	2.18	21.83	45.87
Set III	2.19	21.88	44.95

binding energies are listed in the Table 2 for three different cases. From Table 1 We can see that the mass of σ is somewhat different for these three cases, because it is decided by fitting the deuteron experimental data of 2.22 MeV, the calculated values are shown in Table 2. For no mixing ($\theta^s = 0°$), when m_σ is taken to be 594 MeV, the binding energy of the deuteron is 2.21 MeV. For $\theta^s = -18°$ mixing, when m_σ is taken to be 491 MeV, the binding energy of the deuteron is $2.18 MeV$. For the ideal mixing ($\theta^s = 35.3°$), when m_σ is taken to be 559 MeV, the binding energy of the deuteron is 2.19 MeV. From Table 2, we see that the binding energy for $\Delta\Delta (ST = 03)$ is about 22 MeV, and not changed much with the different

34

mixing. Similarly, for $\Delta\Delta$ ($S T = 30$) state, the binding energy is very stable deeply bound state.

In summary, by introducing the mixing of scalar mesons, we dynamically investigate the structure of possible nonstrange dibaryons in the chiral $S U(3)$ quark model by solving the resonating group method equation. The results show that no matter what kind of mixing is considered, the binding energies of nonstrange dibaryons would become stable if reasonable parameters are used.

Acknowledgments

Supported in part by the National Natural Science Foundation of China (10975068), Scientific Research Foundation of Liaoning Education Department(2009T055), and Doctoral Fund of Ministry of Education of China (201021361110002).

References

1. K. Nakamura *et al.* [Particle Data Group Collaboration], *J. Phys. G* **37**, 075021 (2010).
2. K. Yazaki, *Prog. Theor. Phys. Suppl.* **91**, 146 (1987).
3. F. Wang *et al.*, *Phys. Rev. Lett.* **69**, 2901 (1992);*Phys. Rev. C* **51**, 3411 (1995).
4. H. R. Pang, J. L. Ping, L. Z. Chen and F. Wang, *Chin. Phys. Lett.* **21**,1455 (2004).
5. X. F. Lu, J. L. Ping and F. Wang,*Chin. Phys. Lett.* **20**, 42 (2003).
6. Z. Y. Zhang *et al., Nucl. Phys. A* **625**, 59 (1997).
7. Z. Y. Zhang *et al., Phys. Lett. C* **61**, 065204 (2000).
8. X. Q. Yuan, Z. Y. Zhang, Y. W. Yu, P. N. Shen, *Phys. Rev. C* **60**, 045203 (1999).
9. Q. B. Li, P. N. Shen, *Phys. Rev.C* **62**, 028202 (2000); Q. B. Li, P. N. Shen, Z. Y. Zhang, Y. W. Yu, *Nucl. Phys. A* **683**, 487 (2001).
10. L. R. Dai, *Chin. Phys. Lett.* **22**, 2204 (2005).
11. L. R. Dai, H. Zhang, Y. Fu, Z. Y. Zhang, Y. W. Yu, *Mod. Phys. Lett. A* **23**, 2413 (2008).
12. L. R. Dai, *Chin. Phys. Lett.* **27**, 012102 (2010); **27**, 061301 (2010).
13. L. R. Dai, *Chin. Phys. C* **34**, 1459 (2010).
14. Y. B. Dai and Y. L. Wu, *Eur. Phys. J. C* **39**, S1 (2004).
15. I. T. Obukhovsky and A. M. Kusainov, *Phys. Lett. B* **238**, 142 (1990).
16. A. M. Kusainov, V. G. Neudatchin and I. T. Obukhovsky, *Phys. Rev. C* **44**, 2343 (1991).
17. A. Buchmann, E. Fernandez and K. Yazaki, *Phys. Lett. B* **269**, 35 (1991).
18. E. M. Henley and G. A. Miller, *Phys. Lett. B* **251** 453 (1991).
19. K. Wildermuth and Y.C. Tang, *A Unified Theory of the Nucleus* (Vieweg, Braunschweig, 1977).
20. M. Kamimura, Suppl. Prog. Theor. Phys. **62**, (1977) 236.
21. H. Toki, Z. Phys. A **294**, (1980) 173.
22. M. Oka and K. Yazaki, *Prog. Theor. Phys.* **66**, (1981) 556.
23. U. Straub *et al., Nucl. Phys. A* **483**, (1988) 686.

NET BARYON PRODUCTIONS AND GLUON SATURATION IN THE SPS, RHIC AND LHC ENERGY REGIONS

SHENG-QIN FENG

*Department of Physics, College of Science, China Three Gorges University,
Yichang 443002, China
E-mail: fengsq@ctgu.edu.cn*

The net-baryon number is essentially transported by valence quarks that probe the saturation regime in the target by multiple scattering. The net baryon distributions and nuclear stopping power in the SPS and RHIC energy regions by taking advantage of the gluon saturation model are investigated. A new geometrical scaling variable is used to define the gluon saturation region of central rapidity of central nuclear collisions. Predications for net-baryon rapidity distributions and the mean rapidity loss in central Pb + Pb collisions at LHC are made in this paper.

Keywords: Gluon saturation model; net baryon distributions; color glass condensate.

1. Introduction

In relativistic heavy ion collisions, the features of two distinct and symmetric peaks with respect to rapidity y of net baryon distributions are shown in the CERN (SPS), BNL (RHIC) [1-7] and CERN (LHC) energy regions.

The rapidity separation between the peaks increases with energy and decreasing with increasing mass number, A, reflecting larger baryon stopping for heavier nuclei, as was investigated phenomenologically in the Non-Uniform Flow Model (NUFM).[8-10] In this paper we show how the evolution of the peaks can be linked to saturation physics.

The ideas for the color glass condensate are motivated by HERA data on the gluon distribution function.[11] The gluon density, $xG(x, Q^2)$, rises rapidly as a function of decreasing fractional momentum, x, or increasing resolution, Q. At higher and higher energies, smaller x and larger Q become kinematically accessible. The rapid rise with $\log(1/x)$ was expected in a variety of theoretical works.[12-14] Due to the intrinsic non-linearity of

QCD, gluon showers generate more gluon showers producing an exponential avalanche toward small x. The physical consequence of this exponential growth is that the density of gluons per unit area per unit rapidity of any hadron including nuclei must increase rapidly as x decreases.[15]

Comparing with Ref,[16] we investigate net baryon distributions of central collisions in the SPS and RHIC energy regions by taking advantage of the gluon saturation model with geometric scaling. The net-baryon rapidity distributions and the mean rapidity loss in central Pb + Pb collisions at LHC are predicted in this paper.

2. Gluon saturation model in high energy heavy-ion collisions

The model we considered contains three distinct assumptions, some of which are rather different from those usually included in other gluon saturation models.

(i) We take advantage of the fact that the net-baryon number is essentially transported by valence quarks that probe the saturation regime in the target by multiple scattering. The fast valence quarks in one nucleus scatter in the other nucleus by exchanging soft gluons, leading to their redistribution in rapidity space during the high energy heavy-ion collisions. For symmetric collisions, the contribution of the fragmentation of the valence quarks in the projection is given by the simple formula for the rapidity distribution of interactions with gluon in the target

$$\frac{dN}{dy} = \frac{C}{(2\pi)^2} \int \frac{d^2 p_T}{p_T^2} x_1 q_v(x_1) \varphi(x_2, p_T^2), \qquad (1)$$

here p_T and y are the transverse momentum and rapidity of the produced quark. One important predication of the gluon saturation with geometrical scaling is the geometrical scaling variable p_T^2/Q_s^2, where $Q_s^2(x) = A^{1/3} Q_0^2 x^{-\lambda}$. The fit value $\lambda = 0.2 - 0.3$ agrees with the QCD results.

With the changing of variables

$$x \equiv x_1 \qquad x_2 = x e^{-2y} \qquad p_T^2 = x^2 s e^{-2y}. \qquad (2)$$

(ii) The contribution of valence quarks in the other beam nucleus is added incoherently by changing $y \to -y$. $q_v(x_1)$ is the valence quark distribution of a nucleus, and $\varphi(x_2, p_T)$ is the gluon distribution of another nucleus. The total rapidity distributions of the symmetry interaction systems are the summation of the contributions from the projectile and target, respectively:

$$\frac{dN}{dy}\Big|_{\text{total}} = \frac{dN}{dy}(y) + \frac{dN}{dy}(-y) \tag{3}$$

A scaling variable is introduced:

$$\tau = \ln(s/Q_0^2) - \ln A^{1/3} - 2(1+\lambda)y \tag{4}$$

(iii) By taking $\tau = \ln(1/x)\,|_{x=0.01}$, we can figure out the rapidity region $(0 < y < y_s)$ of gluon saturation as follows:

$$y_s = \frac{1}{1+\lambda}(y_b - \ln A^{1/6}) + \frac{1}{2(1+\lambda)}(\ln\frac{m_n^2}{Q_0^2} - \tau\,|_{x=0.01}). \tag{5}$$

Thus we rewrite the formula as

$$\frac{dN}{dy} = \frac{C}{(2\pi)^2}\int_0^1 \frac{dx}{x}xq_v(x)\varphi(x^{2+\lambda}e^\tau), \tag{6}$$

here $\lambda = 0.2$, $Q_0^2 = 0.04$ GeV^2, $y_b \approx \ln(\sqrt{s}/m_n)$ the beam rapidity with nucleon mass m_n. The gluon distribution is:

$$\varphi(x_2, P_T) = \begin{cases} \text{const} & (x < 0.01) \\ 4\pi\frac{P_T^2}{Q_s(x_2)}\cdot\exp\{-\frac{PT^2}{Q_s(x_2)}\}\cdot(1-x_2)^4 & (x > 0.01) \end{cases} \tag{7}$$

The valence quarks distribution is

$$xq_v(x) = \begin{cases} \propto x^{0.5} & (x < 0.01) \\ \propto (xu_v + xd_v) & (x > 0.01) \end{cases} \tag{8}$$

So the net-baryon distribution which originates from the projectile is

$$\frac{dN}{dy} = \begin{cases} \propto \exp\{\frac{1+\lambda}{2+\lambda}y\} & (0 \leqslant y \leqslant y_s) \\ \propto \int_{0.01}^1 \frac{dx}{x}(xu_v+xd_v)4\pi\frac{P_T^2}{Q_s(x_2)}\exp\{-\frac{PT^2}{Q_s(x_2)}\}(1-x_2)^4 & (y > y_s) \end{cases} \tag{9}$$

The dynamic quark mass effects are considered by substituting the transverse momentum p_T as $\sqrt{(x\sqrt{s}\exp{-y})^2 - m^2}$, here m is the dynamic quark mass. As follow, we will discuss the net-baryon distribution from SPS to LHC energy regions.

3. Net Baryon rapidity distributions in the color glass condensate

The rapidity distributions of net baryon for different energies of relativistic heavy nuclear collisions (Pb + Pb and Au + Au) at SPS and RHIC are given in Figure 1. The integrated net-proton rapidity distributions, scaled by a factor of 2.05 [1-7] to obtain the net-baryon distributions.

The estimated numbers of participants N_{part} are 390, 315, and 357 for $\sqrt{s_{NN}}$ =17.3, 62.4, and 200 GeV, respectively. The solid circles correspond to the experimental result of central collisions[1-7], and real lines are the calculated results from the gluon saturation model with geometric scaling. It is found that our model describes the experimental data of net-baryon distributions very well when we discuss Pb-Pb center collisions at the SPS energy region and Au-Au center collisions at the RHIC energy region.

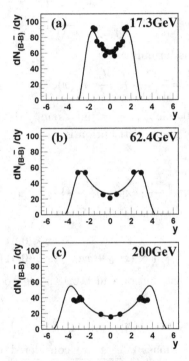

Fig. 1. Net-Baryon distributions of central collisions at SPS $\sqrt{s_{NN}}$=17.3 GeV of Pb-Pb interactions and at RHIC $\sqrt{s_{NN}}$ =62.4 and 200 GeV of Au-Au interactions.[1-7]

By studying the experimental results with gluon saturation model, we can get the values of central rapidity y_s of gluon saturation are 1.06, 2.013 and 3.10 for $\sqrt{s_{NN}}$ =17.3, 62.4, and 200 GeV, respectively.

The net baryon rapidity distribution in central Pb+Pb collisions at LHC energies of $\sqrt{s_{NN}} = 5.52$ TeV is predicted by gluon saturation model with geometric scaling. The theoretical distribution is shown in Fig. 2 for $y_s = 5.86$. The gluon saturation region is larger than that of SPS and

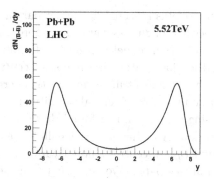

Fig. 2. Rapidity distribution of net baryons in central Pb+Pb collisions at LHC energies of $\sqrt{s_{NN}} = 5.52$ TeV.

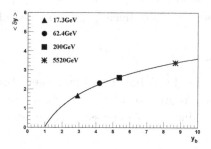

Fig. 3. The dependence of mean rapidity loss $< \delta y >$ on beam rapidity y_b. The ▲, ●, ■ and ∗ are the calculated results by our model for $\sqrt{s_{NN}} =$17.3, 62.4, 200 and 5520 GeV, respectively .

RHIC. It is found that the separation of two symmetric peak of net-baryon is much wider than that of SPS and RHIC.

The mean rapidity loss $< \delta y >= y_b- < y >$ is shown in Fig. 3. We show that the dependence of mean rapidity loss $< \delta y >$ increases on y_b as

$$< \delta y >= 1.548 \cdot \ln(y_b) + 0.036. \tag{10}$$

In Fig. 3, the star $(*)$ is our prediction result of mean rapidity loss $< \delta y >$ for Pb + Pb central collisions at LHC energies of $\sqrt{s_{NN}} = 5.52$ TeV($y_b = 8.68$).

4. Summary and results

A saturation model for net-baryon distributions that successfully describes net-baryon rapidity distributions and their energy dependence is present in this paper. The remarkable feature of geometric scaling predicted by our gluon saturation model is reflected in the net-baryon rapidity distribution, providing a direct test of gluon saturation region and gluon saturation physics. The gluon saturation model is proposed by introducing a scaling variable τ to define the gluon saturation region of central rapidity region of centrally colliding heavy ions at ultra-relativistic energies.

The gluon saturation features of central rapidity at SPS and RHIC can be investigated by using our model. It is found that the values of central rapidity of gluon saturation region increase with colliding energy. The detailed dependence of rapidity (y_s) of central gluon saturation on colliding energy are also investigated in this paper. We also predict the net baryon rapidity distribution in central Pb+Pb collisions at LHC energies of $\sqrt{s_{NN}} = 5.52$ TeV by gluon saturation model with geometric scaling. The gluon saturation region is larger than that of RHIC and SPS , and the separation of two symmetric peak of net-baryon is much wider than that of SPS and RHIC.

Acknowledgments

This work was supported by National Natural Science Foundation of China (10975091), Excellent Youth Foundation of Hubei Scientific Committee (2006ABB036) and Education Commission of Hubei Province of China (Z20081302).

References

1. I. G. Bearden *et al.*, *Phys. Rev. Lett.* **93**, 102301 (2004).
2. I. G. Bearden *et al.*, *Phys. Rev. Lett.* **94**, 162301 (2005).
3. J. L. Klay *et al.*, *Phys. Rev. Lett.* **88**, 102301 (2002).
4. L. Ahle *et al.*, *Phys. Rev. C* **60**, 064901 (1999).
5. J. Barrette *et al.*, *Phys. Rev. C* **62**, 024901 (2000).
6. H. Appelshauser *et al.*, *Phys. Rev. Lett.* **82**, 2471 (1999).
7. I. Arsene *et al.*, BRAHMS Collaboration, *Nucl. Phys. A* **757**, 1 (2005).
8. S. Q. Feng, F. Liu, L. S. Liu, *Phys. Rev. C* **63**, 014901 (2000).
9. S. Q. Feng, X. B. Yuan,Y. F. Shi,*Modern. Phys. Lett. A* **21**, 663 (2006).
10. S. Q. Feng, W. Xiong, *Phys. Rev. C* **77**, 044906 (2008).
11. J. Breitweg *et al.*, *Eur. Phys. J. C* **67**, 609 (1999).
12. L. V. Gribov, E. M. Levin, M. G. Ryskin, *Phys. Rep.* **100**, 1 (1983).
13. A. H. Mueller, J. W. Qiu, *Nucl. Phys. B* **268**, 427 (1986).
14. L. N. Lipatov, *Sov. J. Nucl. Phys.* **23**, 338 (1976).
15. L. D. McLerran, R. Venugopalan, *Phys. Rev. D* **49**, 2233 (1994).
16. Y. Mehtar-Tani, G. Wolschin, *Phys. Rev. Lett.* **102**, 182301 (2009).

PRODUCTION OF HEAVY ISOTOPES WITH COLLISIONS BETWEEN TWO ACTINIDE NUCLIDES

Z. Q. FENG*, G. M. JIN, J. Q. LI

Institute of Modern Physics, Chinese Academy of Sciences, Lanzhou 730000, People's Republic of China
E-mail: fengzhq@impcas.ac.cn

The dynamics of transfer reactions in collisions between two very heavy nuclei ^{238}U$+^{238}$U is studied within the dinuclear system (DNS) model. Collisions between two actinide nuclei form a super-heavy composite system during a very short time, in which a large number of charge and mass transfers may take place. Such reactions have been investigated experimentally as an alternative way for the production of heavy and superheavy nuclei. The role of collision orientation in the production cross sections of heavy nuclides is analyzed systematically. Calculations show that the cross sections decrease drastically with increasing the charged numbers of heavy fragments. The transfer mechanism is favorable to synthesize heavy neutron-rich isotopes, such as nuclei around the subclosure at $N=162$ from No ($Z=102$) to Db ($Z=105$).

Keywords: Transfer reactions; dinuclear system model; superheavy nuclei; heavy neutron-rich isotopes.

The synthesis of superheavy nuclei (SHN) is motivated by searching for the "island of stability", which is predicted theoretically, and has resulted in a number of experimental studies. There are mainly two sorts of reaction mechanism to produce heavy and superheavy nuclei. One is multi-nucleon transfer reactions in collisions between two actinide nuclei.[1,2] The other is fusion-evaporation reactions with a neutron-rich nuclide bombarding a heavy nucleus near shell closure, such as the cold fusion reactions at GSI (Darmstadt, Germany)[3] and the ^{48}Ca bombarding the actinide nuclei at FLNR in Dubna (Russia).[4] However, the decay chains of the nuclei formed by the hot fusion reactions are all neutron-rich nuclides and do not populate presently the known nuclei. Meanwhile, the superheavy isotopes synthesized by the cold fusion and the ^{48}Ca induced reactions are all far away from the doubly magic shell closure beyond ^{208}Pb at the position of protons $Z=114$-126 and neutrons $N=184$. Multi-nucleon transfer reactions in collisions of two actinides might be used to fill the region and examine the influence of the shell effect in the production of heavy isotopes. The production of neutron-rich heavy or superheavy nuclei in low-energy collisions of actinide nuclei was proposed initially by Zagrebaev *et al.*, based on the assumption that the shell effects continue to play a significant role in multi-nucleon transfer reactions.[5]

In this contribution, we use a dinuclear system (DNS) model to investigate the dynamics of the damped collisions of two very heavy nuclei, in which the nucleon transfer is coupled to the relative motion by solving a set of microscopically derived master equations by distinguishing protons and neutrons.[6,7] In order to treat the diffusion process along proton and neutron degrees of freedom in the damped collisions, the distribution probability is obtained by solving a set of master equations numerically in the potential energy surface of the DNS.[8,9]

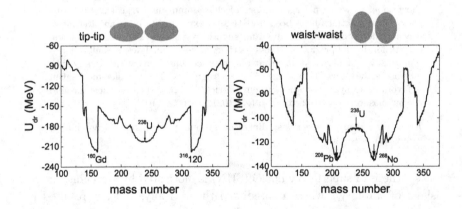

Fig. 1. Driving potentials of the tip-tip and waist-waist collisions in the reaction ^{238}U+^{238}U.

In the relaxation process of the relative motion, the DNS will be excited by the dissipation of the relative kinetic energy. The local excitation energy is determined by the excitation energy of the composite system and the potential energy surface of the DNS. The potential energy surface (PES) of the DNS is given by

$$U(\{\alpha\}) = B(Z_1, N_1) + B(Z_2, N_2) - \left[B(Z, N) + V_{rot}^{CN}(J) \right]$$
$$+ V(\{\alpha\}) \tag{1}$$

with $Z_1 + Z_2 = Z$ and $N_1 + N_2 = N$. Here the symbol $\{\alpha\}$ denotes the sign of the quantities $Z_1, N_1, Z_2, N_2; J, R; \beta_1, \beta_2, \theta_1, \theta_2$. The $B(Z_i, N_i)(i = 1, 2)$ and $B(Z, N)$ are the negative binding energies of the fragment (Z_i, N_i) and the compound nucleus (Z, N), respectively, which are calculated from the liquid drop model, in which the shell and the pairing corrections are included reasonably. The V_{rot}^{CN} is the rotation energy of the compound nucleus. The β_i represent the quadrupole deformations of the two fragments. The θ_i denote the angles between the collision orientations and the symmetry axes of deformed nuclei. In the calculation, the distance R between the centers of the two fragments is chosen to be the value at the touching configuration, in which the DNS is assumed to be formed. So the PES depends on the proton and neutron numbers of the fragments. Shown in Fig. 1 is the calculated PES as functions of mass numbers of fragments for the two cases of the nose-nose and side-side orientations. Dissipation to heavier fragment by nucleon transfer is hindered in the nose-nose collisions and a pocket appears around the nucleus $^{316}120$. The side-side orientation is easily to reach the subclosure ^{268}No, but is also hindered if further dissipation in the high-mass region.

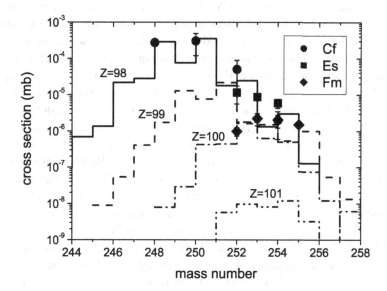

Fig. 2. Calculated mass distributions of the Cf, Es, Fm and Md isotopes at $E_{c.m.}$=800 MeV and compared with the available experimental data.[2]

The cross sections of the primary fragments (Z_1, N_1) after the DNS reaches the relaxation balance are calculated as follows:

$$\sigma_{pr}(Z_1, N_1) = \frac{\pi \hbar^2}{2\mu E_{c.m.}} \sum_{J=0}^{J_{\max}} (2J+1)P(Z_1, N_1, \tau_{int}). \qquad (2)$$

The distribution probability P is obtained by solving a set of master equations numerically.[8,9] The interaction time τ_{int} in the dissipation process is dependent on the incident energy $E_{c.m.}$ in the center-of-mass (c.m.) frame and the angular momentum J has the value of few 10^{-20} s. The survived fragments are the decay products of the primary fragments after emitting the particles and γ rays in competition with fission. The cross sections of the survived fragments are given by

$$\sigma_{sur}(Z_1, N_1) = \frac{\pi \hbar^2}{2\mu E_{c.m.}} \sum_{J=0}^{J_{\max}} (2J+1)P(Z_1, N_1, E_1, \tau_{int})$$
$$\times W_{sur}(E_1, xn, J), \qquad (3)$$

where E_1 is the excitation energy of the fragment (Z_1, N_1). The maximal angular momentum is taken as $J_{\max} = 200$ that includes all partial waves in which the transfer reactions may take place. The survival probability W_{sur} of each fragment can be estimated by using the statistical approach.[7]

Dynamics of the damped collisions was investigated by Zagrebaev and Greiner in detail with a model based on multi-dimensional Langevin equations.[10,11] Larger cross sections in the production of neutron-rich heavy isotopes in collisions of two actinides were pointed out. The collisions of the side-side case need to overcome the higher barrier of the interaction potential in the formation of the DNS. But it is favorable to transfer nucleon by the master equations in the driving potential and form the target-like fragments. Situation is opposite for the nose-nose collisions. So both cases give a similar result in the production of the Bk, Cf, Es and Fm isotopes. Comparison of the calculated mass distributions and the experimental data of the survived fragments at $E_{c.m.}$=800 MeV is shown in Fig. 2. The cross sections decrease drastically with the atomic numbers of fragments. The calculated results are the case of the nose-nose collisions, which have the height of the interaction potential at the touching configuration with the value 713 MeV. In the collisions of such heavy systems, there is no Coulomb barrier in the approaching process of two colliding partners. A number of nucleon transfers take place in the reactions of two actinides owing to the dynamical deformations and the fluctuations of all collective degrees of freedoms in the model of Zagrebaev and Greiner.[10] We assumed the DNS is

formed at the touching configuration in the collisions of two very heavy
nuclei. The nucleon transfer is governed by the driving potential in competition with the quasifission of the DNS because of the collision dynamics.
The larger quasifission rate of such systems results in the DNS quickly decays into two fragments because there is no potential pockets. The inner
excitation energy of the DNS is dissipated from the kinetic energy of the
relative motion overcoming the height of the interaction potential of two
colliding nuclei at the touching configuration. The value of the side-side
orientation is 813.5 MeV. Inclusion of all orientations in the low-energy
damped collisions is important for correctly estimating the cross sections
of the primary and survived fragments.

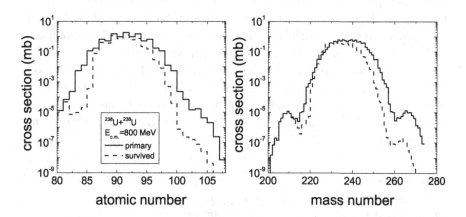

Fig. 3. Cross sections as functions of the charged and mass numbers of the primary
and survived fragments at $E_{c.m.}$=800 MeV, respectively.

Shown in Fig. 3 is the calculated cross sections of the primary and survived fragments by using the Eqs. (2) and (3) as functions of the charged
numbers and mass numbers at the incident energy $E_{c.m.}$=800 MeV, respectively. In the damped collisions, the primary fragments come from a
number of nucleon transfer in the relaxation process of the colliding partners. The giant composite system retains a very short time of several tens
10^{-22} s due to the strong Coulomb repulsion. The cross sections in the production of heavy target-like fragments ($Z > 92$) decrease drastically with
the atomic numbers of the fragments. Therefore, the mechanism of the

low-energy transfer reactions in collisions of two very heavy nuclei is not suitable to synthesize superheavy nuclei (Z >106) because of the smaller cross sections at the level of 1 pb and even below 1 pb. Similar results were also obtained in Ref. 10. However, the production of the survived fragments around the subclosure $N=162$ has a larger cross section. Calculated cross sections as functions of the mass numbers appear a bump near the isotopes of the subclosure. Experimental works for studying the influence of the shell closure in the production of the neutron-rich isotopes should be performed in the near future. It is also a good technique to fill the gap of the new isotopes between the cold fusion and the ^{48}Ca induced reactions.

In summary, the dynamics in collisions of the very heavy system ^{238}U+^{238}U is investigated within the framework of the DNS model. The influence of the collision orientations on the production cross sections of heavy isotopes is discussed systematically. The low-energy transfer reactions in the damped collisions of the actinide nuclei are a good mechanism to produce the neutron-rich heavy isotopes, in which the shell closure plays an important role in the estimation of the cross sections.

This work was supported by the National Natural Science Foundation of China under Grants. 10805061, 10775061 and 10975064, the Special Foundation of the President Fund, the West Doctoral Project of Chinese Academy of Sciences, and the Major State Basic Research Development Program under Grant 2007CB815000.

References

1. E. K. Hulet *et al.*, *Phys. Rev. Lett.* **39**, 385 (1977).
2. M. Schädel *et al.*, *Phys. Rev. Lett.* **41**, 469 (1978).
3. S. Hofmann and G. Münzenberg, *Rev. Mod. Phys.* **72**, 733 (2000); S. Hofmann, *Rep. Prog. Phys.* **61**, 639 (1998).
4. Yu. Ts. Oganessian, *J. Phys. G* **34**, R165 (2007).
5. V. Zagrebaev, Yu. Ts. Oganessian, M. G. Itkis, and W. Greiner, *Phys. Rev. C* **73**, 031602(R) (2006).
6. Z. Q. Feng, G. M. Jin, F. Fu, and J. Q. Li, *Nucl. Phys. A* **771**, 50 (2006).
7. Z. Q. Feng, G. M. Jin, J. Q. Li, and W. Scheid, *Phys. Rev. C* **76**, 044606 (2007); *Nucl. Phys. A* **816**, 33 (2009).
8. Z. Q. Feng, G.M. Jin, and J.Q. Li, *Phys. Rev. C* **80**, 057601 (2009); Phys. Rev. C **80**, 067601 (2009).
9. Z. Q. Feng, G. M. Jin, J. Q. Li, *Nucl. Phys. A* **836**, 82 (2010).
10. V. Zagrebaev and W. Greiner, *J. Phys. G* **34**, 1 (2007); **35**, 125103 (2008).
11. V. Zagrebaev and W. Greiner, *Phys. Rev. C* **78**, 034610 (2008); *Phys. Rev. Lett.* **101**, 122701 (2008).

THE PROJECTED CONFIGURATION
INTERACTION METHOD

ZAO-CHUN GAO* and YONG-SHOU CHEN

China Institute of Atomic Energy, P.O. Box 275-10, Beijing, 102413, China
E-mail: zcgao@ciae.ac.cn

MIHAI HOROI

Department of Physics, Central Michigan University,
Mount Pleasant, Michigan 48859, USA

A new shell model method, called as the Projected Configuration Interaction (PCI), has been established recently. The deformed Slater determinants (SD's) are projected onto good angular momentum and form the PCI basis. Using realistic shell model Hamiltonians, very good approximations, relative to the exact solutions, have been achieved by using a small PCI basis space. Our results have shown that, PCI may provide a promising way to extend the large scale shell model calculations to the heaver nuclear system.

Keywords: Angular momentum projection; configuration interaction; realistic Hamiltonian.

1. Introduction

The full configuration interaction (CI) method using a spherical single particle (s.p.) basis and realistic Hamiltonians, also known as the nuclear shell model, has been very successful in describing various properties of the low-lying states in light and medium nuclei. The realistic Hamiltonians, such as the USD[1,2] in the sd shell, the KB3,[3] FPD6[4] and GXPF1[5] in the pf shell, have provided a very good base to study various nuclear structure problems microscopically.

However, the main limitations of this method are the exploding dimensions with the increase of the number of valence nucleons and with the increase of the valence space. Although, there are continuous improvements to the CI codes[6,7] and computational resources, the exploding CI dimensions significantly restrict the ability to investigate heavy nuclei, especially those that exhibit strong collectivity.

The projection techniques, taking full advantage of the nuclear deformation, and recovering symmetries of the intrinsic states, may be efficient in reducing the dimension of the shell model space. The recent history of projection techniques combined with CI particle-hole configurations includes the projected shell model (PSM)[8,9] and the deformed shell model (DSM) proposed in Ref.[10] Other models using similar techniques includes MONSTER, the family of VAMPIRs,[12] and the quantum Monte Carlo diagonalization (QMCD) method.[13]

Recently, we have proposed a new method of calculating the low-lying states in heavy nuclei, using many particle-hole configurations of spin-projected Slater determinants built on multiple sets of deformed single-particle orbitals.[14,15] This projected configuration interaction (PCI) method takes advantage of the inherent mixing induced by the projected Slater determinants of varying deformations with the many particle-hole mixing typical of CI techniques. Having the deformed basis of Slater determinants chosen, one can use standard spin projection techniques to solve the associated eigenvalues problem. The method proved to be very accurate in $0\hbar\omega$ model spaces, such as sd and pf, where one can easily keep track of different deformed orbitals.

This paper presents an introduction of the PCI method. As an example, the PCI calculation of ^{48}Cr is discussed.

2. The PCI Method

2.1. *The Projected Configuration Interaction method*

The model Hamiltonian used in CI calculations can be written as:

$$H = \sum_i e_i c_i^\dagger c_i + \sum_{i>j,k>l} V_{ijkl} c_i^\dagger c_j^\dagger c_l c_k, \tag{1}$$

where, $c_i^\dagger$ and c_i are creation and annihilation operators of the spherical harmonic oscillator, e_i and V_{ijkl} are one-body and two-body matrix elements.

One can introduce the deformed single particle (s.p.) basis, which can be obtained from a constraint HF solution, or from the Nilsson s.p. Hamiltonian.[11] The deformed s.p. creation operator is given by the following transformation:

$$b_k^\dagger = \sum_i W_{ki} c_i^\dagger, \tag{2}$$

where the matrix elements $W_{ki} = \langle b_k | c_i \rangle$ are real in our calculation. The Slater Determinant (SD) built with the deformed single particle states is given by

$$|\kappa\rangle \equiv |s, \epsilon\rangle \equiv b_{i_1}^\dagger b_{i_2}^\dagger ... b_{i_n}^\dagger |\rangle, \tag{3}$$

where s refers to the Nilsson configuration, indicating the pattern of the occupied orbits, and ϵ is the deformation determined by ϵ_2 and ϵ_4.

The general form of the nuclear wave function is therefore a linear combination of the projected SDs (PSDs),

$$|\Psi_{IM}^\sigma\rangle = \sum_{K\kappa} f_{IK\kappa}^\sigma P_{MK}^I |\kappa\rangle, \tag{4}$$

where $\hat{P}_{MK}^I$ is the angular momentum projection operator. The energies and the wave functions [given in terms of the coefficients $f_{IK\kappa}^\sigma$ in Eq.(4)] are obtained by solving the following eigenvalue equation:

$$\sum_{K'\kappa'} (H_{K\kappa,K'\kappa'}^I - E_I^\sigma N_{K\kappa,K'\kappa'}^I) f_{IK\kappa'}^\sigma = 0, \tag{5}$$

where $H_{K\kappa,K'\kappa'}^I$ and $N_{K\kappa,K'\kappa'}^I$ are the matrix elements of the Hamiltonian and of the norm, respectively

$$H_{K\kappa,K'\kappa'}^I = \langle \kappa | H P_{KK'}^I | \kappa' \rangle, \tag{6}$$
$$N_{K\kappa,K'\kappa'}^I = \langle \kappa | P_{KK'}^I | \kappa' \rangle. \tag{7}$$

2.2. Choice of the PCI Basis

The analysis made in Ref.[14] indicated that one of the most important problems of the PCI method is the proper selection of the PCI basis. As introduced in our previous work,[14,15] the general structure of the PCI basis is

$$\left\{ \begin{array}{ll} 0p - 0h, & np - nh \\ |\kappa_1, 0\rangle, & |\kappa_1, j\rangle, \cdots, \\ |\kappa_2, 0\rangle, & |\kappa_2, j\rangle, \cdots, \\ \cdots\cdots\cdots\cdots\cdots \\ |\kappa_N, 0\rangle, & |\kappa_N, j\rangle, \cdots \end{array} \right\}. \tag{8}$$

$|\kappa_i, 0\rangle$ $(i = 1, ...N)$ is a set of starting states. Assuming that we've found these $|\kappa, 0\rangle$ SDs (We skipped the subscript i to keep notations short.), relative np-nh SDs, $|\kappa, j\rangle$, on top of each $|\kappa, 0\rangle$ are selected by using the formula[14]

$$\Delta E = \frac{1}{2}(E_0 - E_j + \sqrt{(E_0 - E_j)^2 + 4|V|^2}) \geq E_{\text{cut}}, \tag{9}$$

where $E_0 = \langle \kappa, 0|H|\kappa, 0 \rangle$, $E_j = \langle \kappa, j|H|\kappa, j \rangle$ and $V = \langle \kappa, 0|H|\kappa, j \rangle$.

$|\kappa, 0\rangle$ SDs need to be properly chosen. We developed a method of how to find those $|\kappa, 0\rangle$ states.[15] The general idea is to find a set of $|\kappa, 0\rangle$ so that their S_n,

$$S_n = \lambda_1 + \lambda_2 + \cdots + \lambda_n, \tag{10}$$

is as low as possible. Here, $\lambda_1, \lambda_2, ..., \lambda_n$ are eigenvalues of the following generalized eigenvalue equation,

$$\mathbf{A}x = \lambda \mathbf{B}x, \tag{11}$$

with

$$\mathbf{A} = \begin{pmatrix} H_{11} & H_{12} & \dots & H_{1n} \\ H_{21} & H_{22} & \dots & H_{2n} \\ \multicolumn{4}{c}{\dotfill} \\ H_{n1} & H_{n2} & \dots & H_{nn} \end{pmatrix}, \mathbf{B} = \begin{pmatrix} N_{11} & N_{12} & \dots & N_{1n} \\ N_{21} & N_{22} & \dots & N_{2n} \\ \multicolumn{4}{c}{\dotfill} \\ N_{n1} & N_{n2} & \dots & N_{nn} \end{pmatrix},$$

where

$$H_{ij} = \langle i|HP_{MK}^I|j\rangle, N_{ij} = \langle i|P_{MK}^I|j\rangle, \tag{12}$$

$$|i(j)\rangle = |\kappa_{i(j)}, 0\rangle, \text{ if } i(j) = 1, 2, \cdots, n-1;$$

$$|i(j)\rangle = |s, \epsilon\rangle, \text{ if } i(j) = n. \tag{13}$$

Sometimes we calculate the S_n by summing over part of the lowest λ_i's, i.e.,

$$S_n^k = \lambda_1 + \lambda_2 + \cdots + \lambda_k, (1 \le k \le n), \tag{14}$$

and $S_n^n = S_n$ by definition.

3. Calculations of ^{48}Cr

Following the method of Ref.,[15] we performed a PCI calculation for the Yrast states of ^{48}Cr. Here, we adopted the KB3 hamiltonian.[3] For all spins, we set $N = 15$, and $E_{\text{cut}} = 0.5$ keV. The calculated energies, the $E2$ transition energies and the $B(E2)$ values, comparing with the full shell model results, are shown in Fig. 1. The PCI dimensions and the energy differences between PCI and SM are listed in Table 1. It is seen that the PCI results are very close to those of the shell model. All the energy differences are within 100 keV. Such results can be further improved if more Projected SDs are added to the PCI basis.

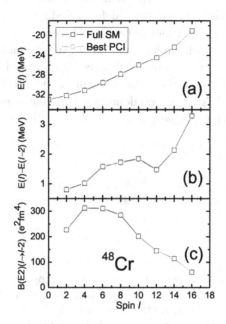

Fig. 1. Results of PCI and full SM. (a) Yrast state energies vs spin. (b) $E2$ transition energy $E(I) - E(I-2)$ vs spin. (c) The $B(E2)$ values with the same wavefunctions as in (a). The effective charges are taken to be $1.5e$ for protons and $0.5e$ for neutrons, which are the same as those used in Refs.[16,17] Results of the full SM are taken from Ref.[16]

Table 1. The PCI dimension and the energy difference between PCI and SM.

Spin	PCI Dimension	$E_{\mathrm{PCI}} - E_{\mathrm{SM}}$(keV)
0	3949	25.75
2	4390	43.49
4	4845	58.85
6	5026	85.36
8	5168	61.58
10	5679	42.24
12	5302	21.22
14	2635	16.01
16	1536	0.33

4. Summary

In this paper, the projected configuration interaction (PCI) method has been briefly introduced. This method takes full advantage of the nuclear deformation and provides effective shell model basis truncation. The selected Slater determinants built with deformed single-particle orbitals are projected onto good angular momentum, and taken as PCI basis states. By adopting the realistic shell model hamiltonian, the calculated PCI energies as well as the $B(E2)$ values can be directly compared with the results of the full shell model calculations. As an example, the PCI calculations of the Yrast states in ^{48}Cr have been done, and excellent approximation with the shell model has been achieved.

Acknowledgments

This work is supported by the NSF of China Contract No. 10775182. Y.S.C. acknowledges support from the MSBRDP of China under Contract No. 2007CB815003.

References

1. B. H. Wildenthal, *Prog. Part. Nucl. Phys.* **11**, 5 (1984).
2. B. A. Brown and B. H. Wildenthal, *Ann. Rev. Nucl. Part. Sci.* **38**, 29 (1988).
3. A. Poves and A. P. Zuker, 1981a, *Phys. Rep.* **71**, 141.
4. W. A. Richter, M. J. van der Merwe, R. E. Julies and B. A. Brown, 1991, *Nucl. Phys. A* **523**, 325.
5. M. Honma, T. Otsuka, B. A. Brown and T. Mizusaki, *Phys. Rev. C* **65**, 061301(R)(2002).
6. E. Caurier and F. Nowacki, *Acta Phys. Pol.* 30, 705 (1999).
7. W. D. M. Rae, NUSHELLX code 2008, http://knollhouse.org/NuShellX.aspx.
8. K. Hara and Y. Sun, *Int. J. Mod. Phys. E* 4, 637 (1995).
9. Y. S. Chen and Z. C. Gao, *Phys. Rev. C* 63, 014314 (2000).
10. S. Mishra, A. Shukla, R. Sahu and V. K. B. Kota, *Phys. Rev. C* 78, 024307 (2008).
11. S. G. Nilsson, *Dan. Mat. Fys. Medd.* 29, 16 (1955).
12. K. W. Schmid, *Prog. Part. Nucl. Phys.* 52, 565 (2004).
13. M. Honma, T. Mizusaki and T. Otsuka, *Phys. Rev. Lett.* 77, 3315 (1996).
14. Zao-Chun Gao and Mihai Horoi, *Phys. Rev. C* **79**, 014311 (2009)
15. Zao-Chun Gao, Mihai Horoi and Y. S. Chen, *Phys. Rev. C* **80**, 034313 (2009)
16. E. Caurier *et al.*, *Phys. Rev. C* **50** 225 (1994).
17. E. Caurier *et al.*, *Phys. Rev. Lett.* **75**, 2466 (1995).

APPLICATIONS OF NILSSON MEAN-FIELD PLUS EXTENDED PAIRING MODEL TO RARE-EARTH NUCLEI

XIN GUAN,* HANG LI, QI TAN, FENG PAN

Department of Physics, Liaoning Normal University, Dalian, 116029, Peoples Republic of China
*E-mail:ggguanxinnn@163.com

J. P. DRAAYER

Department of Physics and Astronomy, Louisiana State University, Baton Rouge, LA 70803-4001, USA

Nilsson mean-field plus the extended pairing model for well-deformed nuclei is applied to describe rare-earth nuclei. Binding energies, even-odd mass differences of Er, Yb, and Hf isotopes are calculated systematically in the model with proton pair excitation frozen approximation. Compared with the corresponding experimental data, the results obtained from the standard pairing model with BCS approximation, and the nearest orbit pairing model, it is shown that the extended pairing model is better than the other two models in description of rare-earth nuclei.

Keywords: Nilsson mean-field; binding energy; odd-even mass difference.

1. Introduction

Besides quadrupole-quadrupole interaction, pairing is another important residual interaction in nuclei.[1,2] Recently, the Nilsson mean-field plus extended pairing model has been introduced to describe well-deformed nuclei.[3] The advantage of the model lies in the fact that not only it is exactly solvable, but also the number of valence nucleon pairs in the model is a conserved quantity. In this talk, we report the extended pairing model application to rare-earth nuclei and comparisons of the results obtained from the standard pairing model and those from the nearest orbit pairing model.

2. The extended pairing model

The standard pairing Hamiltonian for well deformed nuclei is given by

$$\hat{H} = \sum_{j=1}^{p} \epsilon_j n_j - G \sum_{i,j=1}^{p} a_i^\dagger a_j, \tag{1}$$

where p is the total number of Nilsson levels considered, G>0 is the overall pairing strength, ϵ_j is single-particle energies taken from the Nilsson model, $n_j = c_{j\uparrow}^\dagger c_{j\uparrow} + c_{j\downarrow}^\dagger c_{j\downarrow}$ is the fermion number operator for the j-th Nilsson level, and $a_i^\dagger = c_{i\uparrow}^\dagger c_{i\downarrow}^\dagger$ [$a_i = (a_i^\dagger)^\dagger = c_{i\downarrow} c_{i\uparrow}$] are pair creation [annihilation] operators. The up and down arrows in these expressions refer to time-reversed states.

Since to diagonalize the Hamiltonian Eq. (1) in a large Fock subspace is not possible, BCS approximation is often adopted. As an extension of Eq. (1), the extended pairing model was proposed with the Hamiltonian[3]

$$\hat{H} = \sum_{j=1}^{p} \epsilon_j n_j - G \sum_{i,j=1}^{p} a_i^\dagger a_j$$
$$- G\Big(\sum_{\mu=2}^{\infty} \frac{1}{(\mu!)^2} \sum_{i_1 \neq i_2 \neq \cdots \neq i_{2\mu}} a_{i_1}^\dagger a_{i_2}^\dagger \cdots a_{i_\mu}^\dagger a_{i_{\mu+1}} a_{i_{\mu+2}} \cdots a_{i_{2\mu}} \Big), \tag{2}$$

which includes many-pair hopping terms that allow nucleon pairs to simultaneously scatter (hop) between and among different Nilsson levels besides the usual Nilsson mean field and the standard pairing interaction Eq. (1). With this extension in place, the model is exactly solvable.

3. Numerical results and discussions

In the rare earth region, the experimental data for Er, Yb, and Hf isotopes are more abundant, which were then chosen as examples in our study. Binding energies, even-odd mass differences were fitted by the Nilsson mean-field plus extended pairing model and compared to results obtained from the standard pairing and the nearest orbit pairing models. We use mean-square deviation to estimate deviation of binding energies of a chain of isotopes with

$$\sigma = \Big[\sum_{\mu} \big(E_\mu^{\text{th}} - E_\mu^{\text{exp}} \big)^2 / \mathcal{N} \Big]^{\frac{1}{2}}, \tag{3}$$

where E_μ^{th} is theoretical value of the binding energy, E_μ^{exp} is the corresponding experimental value, $\mathcal{N}$ is the total number of nuclei in a chain fitted and the summation runs over all nuclei fitted in a chain. In the fitting, the average binding energy per particle ϵ and the pairing interaction strength G are determined by the corresponding experimental value of the binding energy and the experimental value of the first 0^+ excited state. In the extended pairing model for example, the best fit requires that $\epsilon = -7.475$ MeV for Er isotope, $\epsilon = -7.476$ MeV for Yb isotope, and $\epsilon = -7.71$ MeV for Hf isotope. Specifically, for ^{160}Er, $G = -0.001205$ MeV in the extended pairing model, $G = -0.055$ MeV in the standard pairing model with BCS approximation. In the nearest orbit pairing model, the nearest orbit pairing parameter is $G_{\alpha\beta} = A e^{-B(\epsilon_\alpha - \epsilon_\beta)^2}$ when α and β are the same or the nearest orbits, where ϵ_α is the single particle energy of the orbit α obtained from the Nilsson model.[4] In this case, the parameters $A = -2.33$ MeV, $B = 0.1$ MeV^{-2} in fitting ^{160}Er. Once binding energies are fitted, the odd-even mass difference is calculated by

$$P(A) = E_B(A+1) + E_B(A-1) - 2E_B(A), \tag{4}$$

where $E_B(A)$ is the binding energy of a nucleus with mass number A.

Fig. 1 shows even-odd mass difference of these three chain of isotopes, which is a more sensitive quantity than binding energy in elucidating pairing correlation in nuclei. Though the three models in description of pairing correlation are globally similar, the results show that the extended pairing model with smallest deviations is even much better for every nucleus fitted.

4. Summary

The mean-field plus extended pairing model for well-deformed nuclei is applied to describe rare earth nuclei. With the proton pairing excitation frozen approximation, binding energies, even-odd mass differences of $^{152-169}$Er, $^{154-171}$Yb, $^{158-173}$Hf are calculated, and compared with the corresponding experimental data, the results obtained from the standard pairing model with BCS approximation, and the nearest orbit pairing model. The results show that the extended pairing model is systematically much better than the other two models in fitting these three chains of isotopes as far as binding energies and even-odd mass differences are considered.

Acknowledgments

Support from the U.S. National Science Foundation (PHY-0500291 & OCI-0904874), the Southeastern Universities Research Association, the Natural Science Foundation of China (10775064), the Liaoning Education Depart-

56

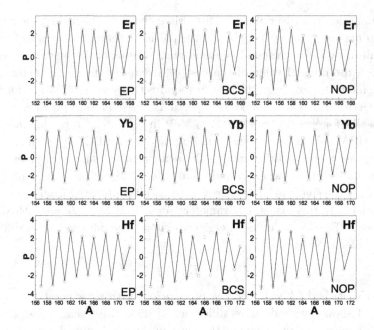

Fig. 1. The theoretical values and the corresponding experimental values of the even-odd mass differences of $^{154-169}$Er, $^{156-171}$Yb, and $^{156-173}$Hf, where P (in MeV) is the even-odd mass difference, A is mass number of the nucleus, the dots show the experimental values, and the theoretical values are connected by the lines. The experimental data are taken from [5].

ment Fund (2007R28), the Doctoral Program Foundation of State Education Ministry of China (20102136110002), and the LSU–LNNU joint research program (9961) is acknowledged.

References

1. J. Bardeen, L. N. Cooper, and J. R. Schrieffer, *Phys. Rev.* **108** (1957) 1175.
2. P. Ring and P. Schuck, *The Nuclear Many-Body Problem* (Springer-Verlag, Berlin, 1980).
3. F. Pan, V. G. Gueorguiev, J. P. Draayer, *Phys. Rev. Lett.* **92** (2004) 112503.
4. H. Molique and J. Dudek, *Phys. Rev. C* **56** (1977) 1795.
5. LBNL-Lund collaboration, Isotope Explorer version 3.0, taken from http://ie.lbl.gov/ensdf/.

COMPLEX SCALING METHOD AND THE RESONANT STATES

JIAN-YOU GUO*, SHOU-WAN CHEN, QUAN LIU

School of Physics and Material Science, Anhui University, Hefei 230039, P.R. China
*E-mail: jianyou@ahu.edu.cn

We have introduced the complex scaling method and its relativistic extension, and demonstrated the utility and applicability of the extended method. We have shown how the resonance states get exposed and the determination of resonance parameters. We have compared the calculated results with the other methods in satisfactory agreements.

Keywords: Complex scaling method; resonant states; relativistic mean field.

1. Introduction

Since the experimental discovery of neutron halo in ^{11}Li,[1] the study of exotic nuclei has become a challenging topic in nuclear structure. To understand the exotic nuclear phenomena, the contribution of the continuum, especially the resonances in the continuum, must be considered for Fermi surface usually closing to the particle continuum in exotic nuclei.[2–5] Thus, the exploration of resonant states is interesting.

There are several techniques for exploring resonant states, such as the R-matrix theory,[6] the extended R-matrix theory,[7] the K-matrix theory,[8] and the conventional scattering theory.[9] Computationally, it is desired to deduce the properties of unbound states from the eigenvalues and eigenfunctions of Hamiltonian for bound states so that the methods developed for bound states can still be used. For this purpose, the bound-state-type methods have been developed, including the real stabilization method (RSM),[10] the analytic continuation in the coupling constant (ACCC) method,[11] and the complex scaling method (CSM).[12]

CSM describes the discrete bound and resonant states on the same footing, and has been widely used to study resonances in atomic and molecular systems[12–14] and atomic nuclei.[15–18] In combination with few-body model, CSM presents good description of the low-lying resonances in a series of light nuclei.[19–26] For example, ^{8}He is regarded as a five-body resonance with CSM.[27] Applying to shell model framework, CSM generates the Berggren basis consisting of bound states, resonance states and continuum scattering states, which induces the Gamow shell model.[28,29] The Gamow shell model is convenient to deal with the multiconfiguration mixing which is important in describing the weakly bound nuclei.[30–34] A recent review can be found in Ref.[18] As the limit of computer capability, Gamow shell model is a tool mainly dedicated to the study of light nuclei. For medium and heavy nuclei, a method of choice is the Hartree-Fock Bogoliubov(HFB). In order to properly treat drip-line nuclei, CSM is used to generate Berggren basis in the HFB calculation.[35–37]

Recently, the relativistic HFB theory has gained the great success in describing the properties for stable nuclei as well as exotic nuclei.[38–41] To extend CSM to the relativistic HFB framework is interesting in checking the resonant states and exotic properties for the nuclei far from the stability line. Many works have been performed on the relativistic extensions of CSM.[42–49] Its application to the relativistic mean field (RMF) is presented in Ref.[50] In this work, we introduce CSM and its relativistic extension, and examine the utility and applicability of the method for realistic nuclei.

2. Complex scaling method

A comprehensive review of the complex scaling method can be found in the literatures.[12,14] Here, only the most essential facts are summarized. The complex scaling is described formally as the complex transformation of the spatial variables:

$$\vec{r} \longrightarrow \vec{r}' = e^{i\theta}\vec{r}, \tag{1}$$

where θ is the so-called rotation angle. To guarantee the wavefunctions of the selected resonances become square integrable,[51–53] the complex scaling operator $U(\theta)$ acting on a single particle wavefunction gives

$$U(\theta)\psi(\vec{r}) = e^{i\frac{3}{2}\theta}\psi(\vec{r}e^{i\theta}) = \psi_\theta(\vec{r}) \tag{2}$$

where the factor $e^{i\frac{3}{2}\theta}$ comes from the three-dimensional volume element. Under $U(\theta)$, the Hamiltonian transforms as

$$H_\theta(\vec{r}) = U(\theta)H(\vec{r})U(\theta)^{-1} \tag{3}$$

The strongly restrictive conditions on the transformation are given with mathematical rigor in the references,[51–53] loosely speaking they amount to the requirement that all quantities in the Hamiltonian are dilation analytic, i.e., the so-called ABC theorem is valid. This theorem states that:

(i) The bound states of H and H_θ are the same;

(ii) The positive-energy spectrum of the original Hamiltonian H is rotated down by an angle of 2θ into the complex-energy plane, exposing a higher Riemann sheet of the resolvent;

(iii) The resonant states of H with eigenvalues $E_{\text{res}} = E_r - i\Gamma/2$ satisfying the condition $|\arg(E_{\text{res}})| < 2\theta$ are also eigenvalues of H_θ and their wavefunctions are square integrable.

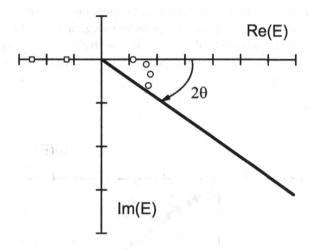

Fig. 1. The complex energy plane of complex scaling Hamiltonian in non-relativistic case (exact solutions). Bound states are labelled as squares on the real energy axis whereas resonant states are shown as circles. The solid corresponds to the rotation continuum.

Since $\psi_\theta(\vec{r})$ for the resonant state is square integrable one can approximate it with the following ansatz:

$$\psi_\theta(\vec{r}) = \sum_{i=1}^{N} c_i(\theta)\, \chi_i(\vec{r}) \tag{4}$$

where $\chi_i(\vec{r})$ $(i = 1, 2, \cdots N)$ are arbitrary known square integrable functions. Putting $\psi_\theta(\vec{r})$ to the equation of motion, one obtains the following inhomogeneous linear system of equations:

$$\sum_{j=1}^{N} [H_{ij}(\theta) - EN_{ij}] c_j(\theta) = 0, \tag{5}$$

where $H_{ij}(\theta) = \int \chi_i^+(\vec{r}) H_\theta(\vec{r})\chi_j(\vec{r})\, d\vec{r}$ and $N_{ij} = \int \chi_i^+(\vec{r}) \chi_j(\vec{r})\, d\vec{r}$. When the function $\chi_i(\vec{r})$ is given, the matrix elements H_{ij} and N_{ij} can be easily calculated, the solution of the equations of motion can be obtained by matrix diagonalization. A demonstrating example is shown in Fig. 1, where the bound states (boxes) that lies on the negative energy axis remains unchanged, the resonance states (circles) located in the fourth quadrant, after the separation from the continuous spectrum, does not vary with θ, the continuum (solid line) along the real positive energy axis rotates clockwise by the angle 2θ. In practical calculation, as the size of the basis adopted is finite, the continuous spectrum gets replaced by a series of discrete points, as shown in Fig. 2.

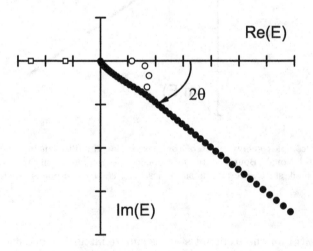

Fig. 2. The complex energy plane of complex scaling Hamiltonian in non-relativistic case (approximating solutions in the finite basis calculations).

The CSM can also be generalized to the relativistic framework, several recent publications can be found in Refs.[44-49] Here, we sketch the relativistic CSM.

In order to study the resonance of relativistic particles, we suppose a spin-1/2 particle with mass M moving in an attracting scalar potential $S(\vec{r})$ and a repulsive vector potential $V(\vec{r})$. Dirac equation reads

$$\{\vec{\alpha} \cdot \vec{p} + V(\vec{r}) + \beta[M + S(\vec{r})]\} \psi(\vec{r}) = \varepsilon\psi(\vec{r}). \tag{6}$$

Here, only the spherically symmetric case is considered, Dirac spinor is written as

$$\psi(\vec{r}) = \begin{pmatrix} f(r)Y_{jm}^l(\vartheta, \varphi) \\ ig(r)Y_{jm}^{\tilde{l}}(\vartheta, \varphi) \end{pmatrix}. \tag{7}$$

Where $\tilde{l} = l \pm 1$, and $'+'$ corresponding to the spin alignment states, $'-'$ corresponding to the spin against states. $Y_{jm}^l(\vartheta, \varphi)$ is the spherical harmonic functions. Putting $\psi(\vec{r})$ into the equation (6), one obtains the radial Dirac equation as

$$\begin{pmatrix} V + S + M & -\frac{d}{dr} - \frac{1}{r} + \frac{\kappa}{r} \\ \frac{d}{dr} + \frac{1}{r} + \frac{\kappa}{r} & V - S - M \end{pmatrix} \begin{pmatrix} f(r) \\ g(r) \end{pmatrix} = \varepsilon \begin{pmatrix} f(r) \\ g(r) \end{pmatrix}. \tag{8}$$

Where $\kappa = \pm (j + 1/2)$. Equation (8) is coupled equation, and is difficult to be solved analytically. Consequently, ones have developed many numerical techniques. For bound states, the basis expansion is one of the commonly used numerical method, and harmonic oscillator wave functions are often used as basis in basis expansion calculation. For unbound states, the numerical methods are also troublesome, especially for many-body system. Therefore, one want to use the bound-state-type approach to the unbound problem. For equation (8), we introduce the complex scaling operator

$$U(\theta) = \begin{pmatrix} e^{i\theta\hat{S}} & 0 \\ 0 & e^{i\theta\hat{S}} \end{pmatrix} \tag{9}$$

with $\hat{S} = \frac{1}{2}\left(r\frac{d}{dr} + \frac{d}{dr}r\right)$. Under $U(\theta)$, Dirac Hamiltonian becomes

$$H_\theta = \begin{pmatrix} \Sigma\left(re^{i\theta}\right) + M & e^{-i\theta}\left(-\frac{d}{dr} - \frac{1}{r} + \frac{\kappa}{r}\right) \\ e^{-i\theta}\left(\frac{d}{dr} + \frac{1}{r} + \frac{\kappa}{r}\right) & \Delta\left(re^{i\theta}\right) - M \end{pmatrix}, \tag{10}$$

where $\Sigma(r) = V(r) + S(r)$ and $\Delta(r) = V(r) - S(r)$. From Ref.[43] the bound states of H are also the eigenstates of H_θ; the resonance poles of H Green operator: $\varepsilon = \varepsilon_r - i\Gamma/2$ is also the eigenstates of H_θ: $\varepsilon_\theta = \varepsilon_r - i\Gamma/2$; the continuous spectrum of H rotates with θ. In the extremely relativistic case, the continuous spectrum rotates around the $\varepsilon-$ origin plane with θ. When $|\varepsilon| \approx M$, the continuous spectrum rotates around the $\varepsilon = \pm M$ with 2θ.

In order to solve the equation (8), the Dirac spinors are expanded as

$$f(r) = \sum_{n=1}^{n_{\max}} f_n(\theta) R_{nl}(r) \; ; \; g(r) = \sum_{\tilde{n}=1}^{\tilde{n}_{\max}} g_{\tilde{n}}(\theta) R_{\tilde{n}\tilde{l}}(r) \,, \tag{11}$$

where $R_{nl}(r)$ are the radial functions of spherical harmonic oscillator potential.

$$R_{nl}(r) = \sqrt{\frac{2(n-1)!}{\Gamma(n+l+1/2)}} \frac{1}{b_0^{3/2}} x^l L_{n-1}^{l+1/2}(x^2) e^{-x^2/2}. \tag{12}$$

Here $x = r/b_0$, b_0 is the oscillator length. $L_n^m(x^2)$ are the associated Laguerre polynomials of order n. The upper limits $n_{\max}$ and $\tilde{n}_{\max}$ are the radial quantum numbers, and label the size of basis adopted here. They are limited by the main shell numbers with $N_{\max} = 2(n_{\max} - 1) + l$ and $\tilde{N}_{\max} = 2(\tilde{n}_{\max} - 1) + \tilde{l}$. The details can be found in the literature.[54]

Inserting the ansatz (11) into Dirac equation (8), and using the orthogonality of wave functions $R_{nl}(r)$ one arrives at a symmetric matrix diagonalization problem,

$$\begin{pmatrix} \Sigma_{n,n'} + M\delta_{n,n'} & T_{n,\tilde{n}'} \\ T_{\tilde{n},n'} & \Delta_{\tilde{n},\tilde{n}'} - M\delta_{\tilde{n},\tilde{n}'} \end{pmatrix} \begin{pmatrix} f_{n'} \\ g_{\tilde{n}'} \end{pmatrix} = \varepsilon \begin{pmatrix} f_n \\ g_{\tilde{n}} \end{pmatrix} \tag{13}$$

of the dimension $n_{max} + \tilde{n}_{max}$. The matrix elements $T_{n,\tilde{n}'}$, $\Sigma_{n,n'}$ and $\Delta_{\tilde{n},\tilde{n}'}$ are given by

$$T_{\tilde{n},n'} = e^{-i\theta} \int_0^\infty R_{\tilde{n}\tilde{l}}(r) \left(\frac{d}{dr} + \frac{1+\kappa}{r} \right) R_{n'l}(r) r^2 dr,$$

$$\Sigma_{n,n'} = \int_0^\infty r^2 dr \left[R_{nl}(r) \Sigma (re^{i\theta}) R_{n'l}(r) \right],$$

$$\Delta_{\tilde{n},\tilde{n}'} = \int_0^\infty r^2 dr \left[R_{\tilde{n}\tilde{l}}(r) \Delta (re^{i\theta}) R_{\tilde{n}'\tilde{l}}(r) \right] \tag{14}$$

$\Sigma_{n,n'}$ and $\Delta_{\tilde{n},\tilde{n}'}$ can be calculated with the Gauss quadrature approximation. The matrix elements $T_{\tilde{n},n'}$ are calculated as

$$T_{\tilde{n},n'} = \frac{1}{b_0 e^{i\theta}} \begin{cases} -\sqrt{\tilde{n}+l+1/2}\delta_{\tilde{n},n'} - \sqrt{\tilde{n}}\delta_{\tilde{n},n'-1}, & \kappa < 0 \\ \sqrt{\tilde{n}+l-1/2}\delta_{\tilde{n},n'} + \sqrt{\tilde{n}-1}\delta_{\tilde{n},n'+1}, & \kappa > 0 \end{cases} \tag{15}$$

With these matrix element $T_{n,\tilde{n}'}$, $V_{n,n'}^+$ and $V_{\tilde{n},\tilde{n}'}^-$, the solution of Dirac equation can be obtained by diagonalizing $H_\theta = \begin{pmatrix} \Sigma_{n,n'} + M\delta_{n,n'} & T_{n,\tilde{n}'} \\ T_{\tilde{n},n'} & \Delta_{\tilde{n},\tilde{n}'} - M\delta_{\tilde{n},\tilde{n}'} \end{pmatrix}$ in an appropriate configuration space. The eigenvalues of H_θ representing bound states or resonant states, do not change with θ, while the eigenvalues representing continuous spectrum rotate. For resonance states, complex energy $\varepsilon = \varepsilon_r - i\Gamma/2$ with ε_r the resonance position and Γ the resonance width. A demonstrating result is shown in Fig. 3, where bound states are labelled as squares on the real energy $\varepsilon-$ axis in the range $+M > \varepsilon > -M$, whereas resonant states are marked as circles. The rotated positive and negative energy continua are the two solid curves.

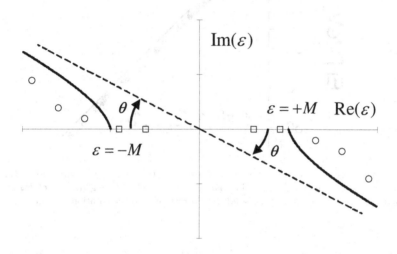

Fig. 3.   The complex energy plane of complex scaling Hamiltonian in relativistic case. Bound states are shown as squares on the real energy $\varepsilon-$ axis whereas resonant states are shown as circles. The rotated positive and negative energy continuum are the two solid curves.

The relativistic CSM is easy to be extended to the framework of relativistic mean field. By using the scalar and vector potentials from the RMF calculations replace the $S(\vec{r})$ and $V(\vec{r})$ in Dirac equation (6), the resonant states from in the RMF-CSM calculations are obtained. The details can be found in the literature.[50]

3. The single particle resonances

In the following, we introduce how the resonances are explored in the RMF-CSM calculations. In Fig. 4, we show the RMF-CSM calculations for the states with the angular momentum $l = 0, 1, 2, \ldots, 14$ in ^{120}Sn. The complex

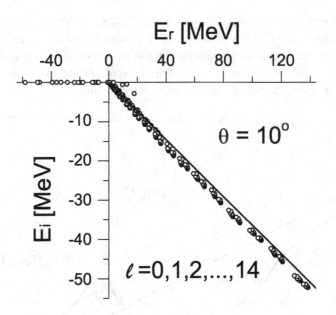

Fig. 4. The eigenvalues of H_θ for these states $l = 0, 1, 2, \ldots, 14$ in the RMF-CSM calculations, where the complex scaling parameter $\theta = 10°$. The eigenvalues of H_θ are labelled in the open circles and the solid line corresponds to the rotation continuum.

rotation angle $\theta = 10°$. All eigenvalues of H_θ are marked as open circles, and the solid line corresponds to the rotation continuum. As the anti-nucleon spectrum is not considered, only the right half of the complex energy plane is plotted in Fig. 4. The position of resonance is marked as $E_\theta = \varepsilon_\theta - M$ instead of the relativistic energy of ε_θ. From Fig. 4, one sees clearly that the eigenvalues of H_θ fall into three regions: the bound states populate on the negative energy axis, the continuous spectrum of H_θ rotates clockwise with the angle θ, and the resonances in the lower half of the complex energy plane located in the sector bound by the new rotated cut line and the positive energy axis get exposed and become isolated. We also note that in the current RMF-CSM calculations, the relativistic effect is weak, the continuous spectrum rotates with 2θ, and is the same as the non-relativistic case.

Next, we demonstrate how the resonant states become exposed by complex rotation. Similar to Ref.,[44] we perform repeated diagonalizations of the eigenvalue problem of H_θ with different θ values. Several illustrated results are depicted in Fig. 5. One sees clearly that the continuous spectrum of H_θ

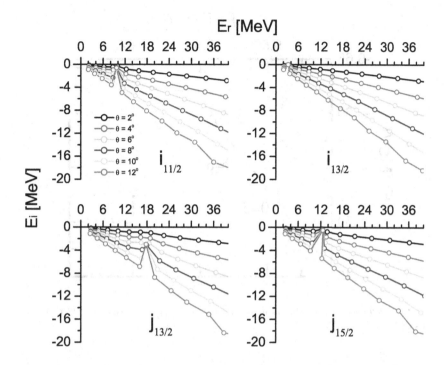

Fig. 5. (Color online) The resonant and continuous spectrum varying with the rotation angle θ for these states $i_{11/2}$, $i_{13/2}$, $j_{13/2}$, and $j_{15/2}$ displayed in the subfigures (a), (b), (c) and (d). The eigenvalues of H_θ are labelled in the open circles.

rotates clockwise with θ, the resonances in the lower half of the complex energy plane located in the sector bound by the new rotated cut line and the positive energy axis get exposed and become isolated. Although the exposure of resonant states relies on the scale of rotation angle, the position of resonant states in the complex energy plane remains almost unchanged with the variation of θ. As the numerical approximation in the present calculations, there will be no eigenenergy that is completely independent of θ. The resonance parameters will move along trajectories in the complex energy plane as a function of θ. The best estimate for the resonance parameters is given by the θ value for which the rate of change with respect to θ is minimal. In Fig. 6, the θ trajectory of resonance parameters varying with θ is plotted. One sees clearly that the resonance parameters moves with θ in the complex energy plane. However, in the vicinity of $\theta = 10°$,

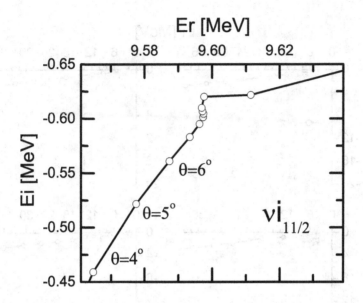

Fig. 6. The θ trajectory of the resonance for the state $i_{11/2}$. θ varies from $4°$ to $14°$ by steps of $1°$.

Table 1. Energies and widths of single neutron resonant states for ^{120}Sn in the RMF-CSM calculations in comparison with the RMF-RSM, RMF-ACCC, and RMF-S calculations. Listed data are presented in the unit MeV.

νl_j	RMF-CSM		RMF-RSM		RMF-ACCC		RMF-S	
	E	Γ	E	Γ	E	Γ	E	Γ
$\nu f_{5/2}$	0.67014	0.01982	0.674	0.030	0.685	0.023	0.688	0.032
$\nu i_{13/2}$	3.26583	0.00403	3.266	0.004	3.262	0.004	3.416	0.005
$\nu i_{11/2}$	9.59732	1.21178	9.559	1.205	9.60	1.11	10.01	1.42
$\nu j_{15/2}$	12.57745	0.99157	12.564	0.973	12.60	0.90	12.97	1.10

the resonance parameters are almost independent of θ, i.e., $\frac{dE_\theta}{d\theta} \approx 0$, which present the resonance position.

In Table I, we compare the results of calculating single neutron resonant states in ^{120}Sn using the four methods. From Table I, one can see clearly the present model provides consistent description for the resonant states with other three methods. The energies and widths from the RMF-CSM calculations are comparable to those from the RMF-RSM,[55] RMF-ACCC,[56] and RMF-S[57,58] calculations. Especially for the low-lying resonant states,

$3f_{5/2}$ and $3i_{13/2}$, the deviations between the present calculations and other three methods are nearly negligible for available resonance parameters. For the higher resonances, $3i_{11/2}$ and $3j_{15/2}$, although there exists a little difference in the widths, the energies obtained are almost same in the four calculations. All these show that the results from the four methods are considerably consistent in describing nuclear resonant states.

4. Summary

In summary, the complex scaling method and its relativistic extension are introduced with the theoretical framework shown. The utility and applicability of the extended method are presented, and how the resonance states get exposed is demonstrated. By investigating the stable behavior of the resonance parameters against change of the complex scaling angle, the resonant states are singled out with the determination of resonance parameters. The resonant states in ^{120}Sn are calculated with satisfactory agreements with those in references.

Acknowledgments

This work was partly supported by the Natural Science Foundation of Anhui Province under Grant No. 11040606M07, the Excellent Talents Cultivation Foundation of Anhui Province under Grant No.2007Z018, and the Education Committee Foundation of Anhui Province under Grant No. KJ2009A129, and the 211 Project of Anhui University.

References

1. I. Tanihata, H. Hamagaki, O. Hashimoto and et al., *Phys. Rev. Lett.* **55**, 2676 (1985).
2. J. Dobaczewski, W. Nazarewicz, T. R. Werner, J.-F. Berger, C. R. Chinn and J. Dechargé, *Phys. Rev. C* **53**, 2809 (1996).
3. W. Pöschl, D. Vretenar, G. A. Lalazissis and P. Ring, *Phys. Rev. Lett.* **79**, 3841 (1997).
4. J. Meng and P. Ring, *Phys. Rev. Lett.* **77**, 3963 (1996) ; *Phys. Rev. Lett.* **80**, 460 (1998).
5. S. G. Zhou, J. Meng, P. Ring and E. G. Zhao, *Phys. Rev. C* **82**, 011301(R) (2010).
6. E. Wigner and L. Eisenbud, *Phys. Rev.* **72**, 29 (1947).
7. G. M. Hale, R. E. Brown, and N. Jarmie, *Phys. Rev. Lett.* **59**, 763 (1987).
8. J. Humblet, B. W. Filippone, and S. E. Koonin, *Phys. Rev. C* **44**, 2530 (1991).
9. J. R. Taylor, *Scattering Theory: The Quantum Theory on Nonrelativistic Collisions* (John Wiley & Sons, New York, 1972).

10. A. U. Hazi and H. S. Taylor, *Phys. Rev. A* **1**, 1109 (1970).
11. V. I. Kukulin, V. M. Krasnopl'sky and J. Horácek, *Theory of Resonances: Principles and Applications* (Kluwer Academic, Dordrecht, 1989).
12. Y. K. Ho, *Phys. Rep.* **99**, 1 (1983).
13. W. P. Reinhardt, *Annu. Rev. Phys. Chem.* **33**, 223 (1982).
14. N. Moiseyev, *Phys. Rep.* **302**, 212 (1998).
15. B. Gyarmati and A. T. Kruppa, *Phys. Rev. C* **34**, 95 (1986).
16. A. T.Kruppa, R. G. Lovas and B. Gyarmati, *Phys. Rev. C* **37**, 383 (1988).
17. Kiyoshi Kato, *J. Phys.: Conf. Ser.* **49**, 73 (2006).
18. N. Michel, W. Nazarewicz, M. Ploszajczak, and T. Vertse, *J. Phys. G: Nucl. Part. Phys.* **36**, 013101 (2009).
19. T. Myo, K. Katō, S. Aoyama, K. Ikeda, *Phys. Rev. C* **63**, 054313 (2001).
20. T. Myo, S. Aoyama, K. Katō, K. Ikeda, *Phys. Lett. B* **576** (2003) 281.
21. K. Yoshida and K. Hagino, *Phys. Rev. C* **72**, 064311 (2005).
22. K. Arai, *Phys. Rev. C* **74**, 064311 (2006).
23. A. T. Kruppa, R. Suzuki and K. Kato, *Phys. Rev. C* **75**, 044602 (2007).
24. E. Garrido, D.V. Fedorov, H. O. U. Fynbo and A. S. Jensen, *Phys. Lett. B* **648**, 274 (2007).
25. T. Myo, R. Ando and K. Katō, *Phys. Rev. C* **80**, 014315 (2009).
26. T. Matsumoto, K. Katō and M. Yahiro, *Phys. Rev. C* **82**, 051602(R) (2010).
27. T. Myo, R. Ando and K. Katō, *Phys. Lett. B* **691**, 150 (2010).
28. N. Michel, W. Nazarewicz, M. Ploszajczak and K. Bennaceur, *Phys. Rev. Lett.* **89**, 042502 (2002).
29. R. Id Betan, R. J. Liotta, N. Sandulescu and T. Vertse, *Phys. Rev. Lett.* **89**, 042501 (2002).
30. N. Michel, W. Nazarewicz and M. Ploszajczak, *Phys. Rev. C* **70**, 064313 (2004).
31. G. Hagen, M. Hjorth-Jensen and N. Michel, *Phys. Rev. C* **73**, 064307 (2006).
32. J. Dobaczewski, N. Michel, W. Nazarewicz and etal., *Prog. Part. Nucl. Phys.* **59**, 432 (2007).
33. N. Michel, W. Nazarewicz and M. Ploszajczak, *Phys. Rev. C* **82**, 044315 (2010).
34. N. Michel, W. Nazarewicz, J. Okolowicz and M. Ploszajczak, *J. Phys. G: Nucl. Part. Phys.* **37**, 064042 (2010).
35. A. T. Kruppa, P. H. Heenen, H. Flocard and R. J. Liotta, *Phys. Rev. Lett.* **79**, 2217 (1997).
36. A. T. Kruppa, P. H. Heenen and R. J. Liotta, *Phys. Rev. C* **63**, 044324 (2001).
37. N. Michel, K. Matsuyanagi and M. Stoitsov, *Phys. Rev. C* **78**, 044319 (2008).
38. D. Vretenar, A. V. Afanasjev, G. A. Lalazissis and P. Ring, *Phys. Rep.* **409**, 101 (2005).
39. J. Meng, H. Toki, S. G. Zhou, S. Q. Zhang, W. H. Long and L. S. Geng, *Prog. Part. Nucl. Phys.* **57**, 470 (2006).
40. G.A. Lalazissis, *Prog. Part. Nucl. Phys.* **59**, 277 (2007).
41. T. Nikšić, D. Vretenar, P. Ring, *Prog. Part. Nucl. Phys*, (2011), in press
42. R. A. Weder, *J. Math. Phys.* **15**, 20 (1974).

43. P. Seba, *Lett. Math. Phys.* **16**, 51 (1988).
44. I. A. Ivanov and Y. K. Ho, *Phys. Rev. A* **69**, 023407 (2004).
45. G. Pestka, M. Bylicki and J. Karwowski, *J. Phys. B: At. Mol. Opt. Phys.* **39**, 2979 (2006).
46. A. D. Alhaidari, *Phys. Rev. A* **75**, 042707 (2007).
47. M. Bylicki, G. Pestka, and J. Karwowski, *Phys. Rev. A* **77**, 044501 (2008).
48. Jian-You Guo, Miao Yu, Jing Wang, Bao-Mei Yao and Peng Jiao, *Comput. Phys. Commun.* **181**, 550 (2010).
49. Jian-You Guo, Jing Wang, Bao-Mei Yao and Peng Jiao, *Int. J. Mod. Phys. E* **19**, 1357 (2010).
50. Jian-You Guo, Xiang-Zheng Fang, Peng Jiao, Jing Wang and Bao-Mei Yao, *Phys. Rev. C* **82**, 034318 (2010).
51. J. Aguilar and J. M. Combes, *Commun. Math. Phys.* **22**, 269 (1971).
52. E. Balslev and J. M. Combes, *Commun. Math. Phys.* **22**, 280 (1971).
53. B. Simon, *Commun. Math. Phys.* **27**, 1 (1972).
54. Y. K. Gambhir, P. Ring and A. Thimet, *Ann. Phys. (NY)* **198**, 132 (1990).
55. L. Zhang, S. G. Zhou, J. Meng and E. G. Zhao, *Phys. Rev. C* **77**, 014312 (2008).
56. S. S. Zhang, J. Meng, S. G. Zhou and G. C. Hillhouse, *Phys. Rev. C* **70**, 034308 (2004).
57. L. G. Cao and Z. Y. Ma, *Phys. Rev. C* **66**, 024311 (2002).
58. N. Sandulescu, L. S. Geng, H. Toki and G. C. Hillhouse, *Phys. Rev. C* **68**, 054323 (2003).

PROBING THE EQUATION OF STATE BY DEEP SUB-BARRIER FUSION REACTIONS

HONG-JUN HAO* and JUN-LONG TIAN**

*College of Physics and Electrical Engineering, Anyang Normal University,
Anyang, Henan 455000, China*
*E-mail: hhj@aynu.edu.cn
**E-mail: tianjunlong@gmail.com

In the framework of Skyrme energy-density functional approach together with the extended semi-classical Thomas-Fermi method, the nuclear equation of state (EOS) is investigated using the nucleus-nucleus potential. The potentials of $^{58}Ni+^{58}Ni$ and $^{32}S+^{89}Y$ are studied in the point of view that the energy at touching point is equal to the threshold energy for hindrance. It is found that MSK3-MSK5, V110 are suitable Skyrme parameters sets and the incompressibility values are in the range of $K = 231$-234 MeV.

Keywords: Skyrme energy-density functional; entrance channel potential; equation of state; incompressibility.

The equation of state (EOS) for the nuclear matter is usually defined as a relation between energy per nucleon E and nucleon density ρ. From the properties of nuclear matter, such as the pressure and incompressibility, the EOS can be obtained. Although the study of giant dipole resonances[1,2] in the ^{208}Pb generally yields the EOS with incompressibility $K = 240 \pm 20$ MeV, this value is about 10% smaller[3,4] when Sn experimental data are used. Therefore, an appropriate choice of the nuclear equation of state is still far from settlement.

In this talk, we study the nuclear EOS by examining fusion reactions at sub-barrier energies. Recent years, a new phenomenon[5-7] of unexpected steep falloff of cross section in deep sub-barrier fusion reactions, referred to as hindrance, has been found bellow certain threshold energies. The hindrance reveals[8,9] that the form of the nucleus-nucleus (NN) potential should have a shallow pocket in the inner part of the barrier. Ichikawa *et al.*,[10] found that the threshold energy E_s for hindrance observed is strongly correlated with the potential energy at the touching point of the colliding nuclei. Based on these works, we adopt a point of view that the touching potential of two colliding nucleus V_{touch} is equal to the threshold energy E_s, to study the nuclear EOS and the Skyrme parameters sets.

We calculate the NN potential, by using the Skyrme energy density functional together with the semi-classical extended Thomas-Fermi (ETF) approach.[11,12] Forty nine Skyrme interactions sets, which associate with different types of EOS, will be used.

The interaction potential $V(R)$ between two ions in the model can be written as

$$V(R) = \int \left\{ \mathcal{H}[\rho_1(\mathbf{r}) + \rho_2(\mathbf{r} - \mathbf{R})] - \mathcal{H}[\rho_1(\mathbf{r})] - \mathcal{H}[\rho_2(\mathbf{r})] \right\} d\mathbf{r}, \quad (1)$$

where R is the center-to-center distance of the two interacting nuclei, while ρ_1 and ρ_2 are the frozen densities of the projectile and the target, respectively. $\mathcal{H}$ is the energy density functional, which consists of the kinetic energy, nuclear interaction (Skyrme force) and Coulomb interaction energy parts

$$\mathcal{H} = \frac{\hbar^2}{2m}[\tau_n(\mathbf{r}) + \tau_p(\mathbf{r})] + H_{\text{Sky}}(\mathbf{r}) + H_{\text{Coul}}(\mathbf{r}). \quad (2)$$

Where τ_n and τ_p are kinetic energies for neutrons and protons, respectively. The kinetic energy is expressed by using the extended Thomas-Fermi approach including 2nd- and 4th-order in the spatial derivatives. So, the potential is a functional of protons and neutrons densities. The density distributions of nucleus are taken as spherical symmetric Fermi functions,

$$\rho_i(\mathbf{r}) = \rho_{i0}[1 + \exp(\frac{r - R_{i0}}{a_i})]^{-1}, \quad i = n, p. \quad (3)$$

Here ρ_{i0} is central density. R_{p0}, a_p, R_{n0} and a_n are the radius and diffuseness for proton and neutron density distributions, respectively.

Firstly, we reexamine the hypothesis that the potential energy at the touching point equals to the threshold incident energy of fusion hindrance. The touching point is the summation of radii of the projectile and the target, $r_{\text{touch}} = R_1 + R_2$. The nuclear radius is adopted as $R = r_0 A^{1/3}$, where $r_0 = 1.2$ fm. The touching energies V_{touch} for 15 systems are calculated with the Skyrme force SkM*. Figure 1 shows the differences between V_{touch} and E_s. It is found that the threshold energy E_s approximately equals to the touching energies V_{touch} in the admissible error bound. For some systems, the lager errors may be attributed to the improper Skyrme forces.

Then we study how the EOS affects the NN potential. In Fig. 2, four Skyrme forces with different incompressibility values are used to calculate

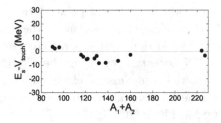

Fig. 1. The differences between V_{touch} and E_s for 15 systems: ^{28}Si+^{62}Ni, ^{16}O+^{76}Ge, ^{16}Si+^{64}Ni, ^{48}Ca+^{48}Ca, ^{58}Ni+^{58}Ni, ^{58}Ni+^{60}Ni, ^{32}S+^{89}Y, ^{58}Ni+^{64}Ni, ^{64}Ni+^{64}Ni, ^{40}Ca+^{90}Zr, ^{58}Ni+^{74}Ge, ^{64}Ni+^{74}Ge, ^{60}Ni+^{89}Y, ^{16}O+^{144}Sm, ^{16}O+^{208}Pb, ^{19}F+^{208}Pb. $A_1 + A_2$ is the total mass number of the system.

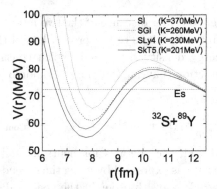

Fig. 2. The NN potential as a function of the center-of-center distance for ^{32}S+^{89}Y system. Four Skyrme forces, with different incompressibility values in EOS, are used to calculated the barrier. The horizontal line is the threshold energy E_s.[7]

the NN potential of ^{32}S+^{89}Y system. The 4 Skyrme forces are SI ($K = 370$ MeV), SGI ($K = 260$ MeV), SLy4 ($K = 230$ MeV), and SkT5 ($K = 201$ MeV). It is found that the incompressibility, so EOS, affects the profile of potential drastically.

In the point of view that the potential at touching point is equal to the threshold energy, we study the nuclear EOS as well as the Skyrme parameters set. Forty-nine Skyrme force parameters (SI, SII, SIII, FItB, SGI, SGII, SkM*, SKT1-SKT9, SkP, Es, Gs, Rs, Zs, SkMp, SkSC4, SkI1, SkI5, SLy0-SLy10, SkM1, SLy230a, SkX, SkO, MSK1-MSK6, V110, BSK1, BSk8, LNS) have been used to calculate the ^{58}Ni+^{58}Ni potential (Fig. 3). The touching energies in the range of $E_s \pm 1\sigma$ error are considered suitable. The following Skyrme force parameters sets are satisfied: MSK1-MSK6, SKT1-SKT3, SKT6-SKT9, V110, SKP, SII, SLy6, SKMP. Same methods are applied to ^{32}S+^{89}Y(no figure), and the suitable parameters sets are SLy4, SLy10, SGI,

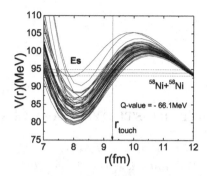

Fig. 3. The NN potential as a function of the center-of-center distance for ^{58}Ni+^{58}Ni system, calculated with 49 Skyrme forces. The horizontal lines are the threshold energy E_s and its 1σ error bar.[7] The vertical line denotes the touching position r_{touch}. The reaction Q-value of the system is also given.

SkO, Es, SKSC4, SkI1, MSK3, MSK4, MSK5 and V110. Finally, we conclude that MSK3-MSK5, V110 are suitable Skyrme parameters sets and the incompressibility values are in the range of $K = 231 - 234$ MeV.

Acknowledgments

This work was supported by the National Natural Science Foundation of China under Grant Nos. 11005003, 11005002, 11047108 and the Natural Science Foundation of Henan Educational Committee Nos. 2010B140001, 2011A140001.

References

1. D. H. Youngblood, H. L. Clark and Y. W. Lui, *Phys. Rev. Lett.* **82**, 691 (1999).
2. D. Vretenar, T. Niksic and P. Ring, *Phys. Rev. C* **68**, 024310 (2003).
3. J. Li, G. Colo and J. Meng, *Phys. Rev. C* **78**, 064304 (2008).
4. J. Piekarewicz, *Phys. Rev. C* **76**, 031301 (2007); *J. Phys. G* **37**, 064038 (2010).
5. C. L. Jiang, K. E. Rehm, R. Janssens, *et al.*, *Phys. Rev. Lett.* **93**, 012701 (2004).
6. C. L. Jiang, B. B. Back, H. Esbensen, *et al.*, *Phys. Rev. B* **640**, 18 (2006).
7. C. L. Jiang, B. B. Back, H. Esbensen, R. V. F. Janssens and K. E. Rehm, *Phys. Rev. C* **73**, 014613 (2006).
8. C. H. Dasso and G. Pollarolo, *Phys. Rev. C* **68**, 054604 (2003).
9. S. Misicu and H. Esbensen, *Phys. Rev. Lett.* **96**, 112701 (2006).
10. T. Ichikawa, K. Hagino and A. Iwamoto, *Phys. Rev. C* **75**, 064612 (2007).
11. J. Bartel and K. Bencheikh, *Eur. Phys. J. A* **14**, 179 (2002).
12. M. Liu, N. Wang, Z. X. Li, X. Z. Wu and E. G. Zhao, *Nucl. Phys. A* **768**, 80 (2006).

DOUBLET STRUCTURE STUDY IN A ~105 MASS REGION

C. Y. HE*, X. G. WU*, Y. ZHENG, B. ZHANG, X. Q. LI, G. S. LI, B. B. YU

China Institute of Atomic Energy, Beijing, 102413, China
*E-mail: chuangye.he@gmail.com *E-mail: wxg@ciae.ac.cn

L. H. ZHU

Beijing University of Aeronautics and Astronautics, Beijing, 100191, China
E-mail: zhulh@buaa.edu.cn

Chiral bands in the A ~105 neutron deficient nuclei have been extensively studied. Doublet structures similar to chiral bands have been found in the nuclei around ^{104}Rh. Apart from those three signatures suggested for the chiral bands in ^{104}Rh, alignments and moment of inertia have been discussed for the doublet structures together with the lifetime measurement.

Keywords: High spin states; chiral bands; momentum of inertia; alignment.

1. Introduction

Chiral symmetry breaking has been predicted recently to exist in odd-odd nuclei with triaxial deformations.[1] In the nuclei with γ ~30°, the angular momentum of high-j low Ω orbital valence proton/neutron, high-j high Ω orbital valence neutron/proton and the collective core are favors to perpendicular coupling to each other. Thereby, the qualifications for chiral bands are coming into being. It has been suggested that chiral bands might be observed in the nuclei of transitional A ~105 mass region. In this paper, we will report the recent progress of the chirality study at HI-13 tandem accelerator of China Institute of Atomic Energy in this region.

2. High-spin spectroscopic results at HI-13 tandem accelerator

In collaborations with Tsinghua University and Shandong University, the high spin states in ^{106,112}In, ^{104,106,108}Ag and ^{98}Tc were investigated by using fusion evaporation reactions. An array including 13 Compton-suppressed HPGe detectors and 1 planar HPGe detector was used for the γ

rays detection. These detectors were placed at angles ±30°, ±36°, 120° and 90° relative to the beam line. Coincidence data were recorded when two or more HPGe detectors were fired in the prompt time. The γ-γ coincidence data were sorted offline into symmetry matrix and DCO matrix. Some of the preliminary results were published[2-6] and the experimental details were also described thereinto. Apart from ^{104}Ag, the level schemas of these nuclei listed above were all considerably updated compared with the previous results.[7-11]

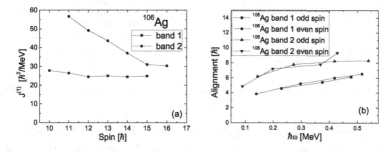

Fig. 1. (a) MOI for band 1 and 2 in ^{106}Ag. (b) Alignments of band 1 and 2 in ^{106}Ag.

3. Discussion

In ^{106,108}Ag and ^{98}Tc, doublet bands similar to the chiral bands in ^{104}Rh [12] have been found. In the recent publications,[3-5] the properties of the doublets bands in these three nuclei were found to be reasonably in accordance with the predicted experimental fingerprints [12] for chirality. (i) The energies are near degeneracy for states with same spin in the two bands. (ii) The signatures are smoothly going with spin increasing. (iii) Both partners have the same characteristic staggering of the $B(M1)/B(E2)$ ratios for in-band transitions. However, recent publications[13,14] reported another two necessary conditions for chirality, besides with these three key criteria listed above. One is that the quasiparticle alignments for the two partners of the doublet bands should be very similar, the other is that their kinematic moments of inertia (MOI) are supposed to be close to each other. It is obvious from Fig. 1, 2 and 3 that both the MOI and the quasiparticle alignments of the doublet bands in ^{98}Tc are very similar except of ^{106,108}Ag. The resemblance of the MOI values explicitly indicates a similarity of shape of the collective mean field whereas the similar quasiparticle angular momentum indicates a similarity of the intrinsic quasiparticle structure responsible for the two bands. It is suggested that the doublet bands in ^{98}Tc are good candidate of chiral bands, other than those in ^{106}Ag and ^{108}Ag.

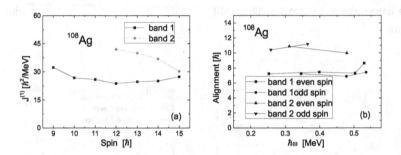

Fig. 2.   (a) MOI for band 1 and 2 in ^{108}Ag. (b) Alignments of band 1 and 2 in ^{108}Ag.

It is apparent in Fig. 1 and 2 that the doublet bands in ^{106}Ag and ^{108}Ag have very similar characteristics. The moment of inertia for both partners in ^{106}Ag and ^{108}Ag are getting closer with increasing spin, and their alignments in both nuclei have about $2\hbar$ difference. The same properties in ^{106}Ag and ^{108}Ag indicate their similar structures in these two nuclei. In order to examine the properties of the doublets bands in ^{106}Ag and ^{108}Ag, we have measured the lifetime for the doublets states in ^{106}Ag using Doppler-shift attenuation method. The preliminary results [15] show that there is a large difference for the experimental $B(E2)$ values between band 1 and 2 in ^{106}Ag. The measured in-band $B(E2)$ values in band 2 are greater than these in band 1. Therefore, a possible conclusion could be drawn that the deformation in band 1 is smaller than that in band 2. In Fig. 1, the moment of inertia in band 2 is greater than band 1, thus confirms the conclusion from lifetime measurement.

4. Summary

Doublet bands in the $A \sim 105$ neutron deficient nuclei have been extensively studied at HI-13 tandem accelerator of China Institute of Atomic Energy. Doublet structures have been found in ^{106,108}Ag and ^{98}Tc. By comparison of the moment of inertia and of alignments for both partners in these nuclei, ^{98}Tc could be a good candidate of chiral bands. Lifetime measurement for

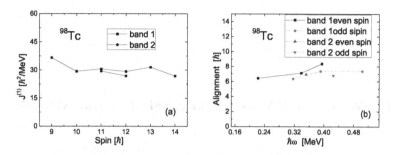

Fig. 3. (a) MOI for band 1 and 2 in ^{98}Tc. (b) Alignments of band 1 and 2 in ^{98}Tc.

the doublet bands in ^{106}Ag indicates that the shape of band 2 is probably more deformed than band 1.

Acknowledgments

The authors are grateful to the crew of HI-13 tandem accelerator in the China Institute of Atomic Energy for steady operation of the accelerator and to Dr. Q. W. Fan for preparing the target. This work is supported by the Major State Basic Research Development Program (2007CB815000) and by the National Natural Science Foundation of China (11075214,10927507, 10975191, 10675171, 10105015, 10375092, 10575133.) etc.

References

1. S. Frauendorf and J. Meng, *Nucl. Phys. A* **617**, 131 (1997).
2. C. Y. He *et al.*, *Eur. Phys. J. A* **46**, 1 (2010).
3. C. Y. He *et al.*, *High. Phys. And Nucl. Phys.* **30(S2)**, 166 (2006).
4. C. Y. He *et al.*, *Atomic Energy Sci. Technol.* **45**, 129 (2011).
5. H. B. Ding *et al.*, *Chin. Phys. Lett.* **27**, 072501 (2010).
6. C. Y. He *et al.*, *Phys. Rev. C* **83**, 024309 (2011).
7. D. Seweryniak *et al.*, *Nucl. Phys. A* **589**, 175 (1995).
8. M. Eibert *et al.*, *J. Phys. G* **2**, L203 (1976).
9. D. Jerrestam *et al.*, *Nucl. Phys. A* **577**, 786 (1994).
10. F. R.Espinoza-Quinones *et al.*, *Phys. Rev. C* **52**, 104 (1995).
11. A. M. Bizzeti-Sona *et al.*, *Phys. Rev. C* **36**, 2330 (1987).
12. C. Vamen *et al.*, *Phys. Rev. Lett.* **92**, 032501 (2004).
13. C. M. Petrache *et al.*, *Phys. Rev. Lett.* **96**, 112502 (2006).
14. P. Joshi *et al.*, *Phys. Rev. Lett.* **98**, 102501 (2007).
15. Y. Zheng *et al.*, *Eur. Phys. J. A*, submitted.

ROTATIONAL BANDS IN TRANSFERMIUM NUCLEI

X. T. HE

College of Material Science and Technology,
Nanjing University of Aeronautics and Astronautics, Nanjing, 210016, China
E-mail: hext@nuaa.edu.cn

Rotational bands of ^{250}Fm, ^{252,253,254}No and ^{251}Md are studied by the cranked shell model with particle-number conserving treatment for monopole and quadrupole pairing correlations. Observed bands are reproduced very well by theoretical results. Backbendings of kinematic moment of inertia $\Im^{(1)}$ and the alignment of high-j orbital are investigated.

Keywords: Rotational band; transfermium nuclei; high-j orbital.

1. Introduction

Exploration of the region of superheavy elements (SHE) has been one of the most fascinating topics in modern physics. Great progress has been made in searching for the superheavy element. Up to date, identifications of elements $Z = 110 \sim 118$ have been claimed.[1-3] However, due to the extremely low production cross-sections, these experiments can rarely reveal the detailed spectroscopic information. The heaviest systems which are accessible in present in-beam experiment are the transfermium nuclei. The strongly downsloping orbitals originating from the spherical subshells active in the vicinity of the predicted shell closures come close to the Fermi surface of transfermium nuclei. Therefore, such research could provide an indirect information of the single particle levels which are relevant to the location and properties of the next shell closures above lead. Rotational bands have been observed in ^{250}Fm, ^{252,253,254}No,[4-11] ^{251}Md[12] and ^{255}Lr.[13] High-K structures are also observed in ^{252,254}No,[14,15] ^{250}Fm,[16] ^{256}Rf,[17] ^{257}Rf[18] and ^{255}Lr.[19] Rotational bands in ^{253}No, ^{251}Md and in ^{247,249}Cm, ^{249}Cf have been studied by the cranked shell model (CSM) with the particle-number conserving (PNC) treatment for monopole and quadrupole pairing correlations.[20,21] In this paper, we will investigate the rotational bands of ^{250}Fm, ^{252,253,254}No, and ^{251}Md by PNC-CSM method.

2. PNC method

The cranked shell model Hamiltonian with pairing reads:

$$H_{\text{CSM}} = H_{\text{SP}} - \omega J_x + H_{\text{P}} = H_0 + H_{\text{P}}, \qquad (1)$$

where $H_0 = \sum_i (h_{\text{Nilsson}} - \omega j_x)_i$, h_{Nilsson} the Nilsson Hamiltonian, $-\omega j_x$ the Coriolis force. H_{P} is the pairing including monopole and quadrupole pairing correlations. We diagonalize H_{CSM} in a sufficiently large cranked many-particle configuration (CMPC) space to obtain the yrast and low-lying eigenstates. The eigenstate of H_{CSM} is expressed as:

$$|\psi\rangle = \sum_i C_i |i\rangle, \qquad (2)$$

where $|i\rangle$ denotes an occupation of particles in the cranked Nilsson orbitals and C_i is the corresponding probability amplitude. The angular momentum alignment $\langle J_x \rangle$ of the state $|\psi\rangle$ is given by:

$$\langle\psi| J_x |\psi\rangle = \sum_i |C_i|^2 \langle i| J_x |i\rangle + 2 \sum_{i<j} C_i^* C_j \langle i| J_x |j\rangle. \qquad (3)$$

3. Results and discussions

The Nilsson parameters κ, μ are taken from Ref.[22] we take the quadrupole deformation parameter of $\varepsilon_2 = 0.29(0.30)$ for ^{250}Fm, ^{252,253,254}No (^{251}Md). The hexadecapole deformation parameter of $\varepsilon_4 = 0.02$ for all nuclei. The effective pairing strengths are determined by fitting the values of $\Im^{(1)}$ for ^{252}No in the frequency range $\hbar\omega = 0.10 \sim 0.25$ MeV. G_0 for monopole and G_2 for quadrupole pairing correlations in units of MeV are $G_{0p}=0.45$, $G_{0n}=0.35$, $G_{2p}=0.02$, $G_{2n}=0.02$.

Our calculated Nilsson level sequence near the Fermi surface of ^{253}No are shown in Fig. 1. The comparisons of kinematic moments of inertia $\Im^{(1)}$ between the experimental and calculated rotational bands of ^{252,254}No, ^{250}Fm and of ^{253}No, ^{251}Md are presented in Figs. 2 and 3, respectively. The observed experimental data are reproduced very well by theory. There is no backbending occurring in the whole observed rotational frequency. According to the calculation, backbendings would take place at $\hbar\omega \approx 0.275$ MeV for bands of ^{252,253,254}No. This due to the rapidly aligned angular momentum of proton $1j_{15/2}$ [770]1/2 pairs and neutron $2h_{11/2}$ [761]1/2, $1j_{15/2}$ [734]9/2 pairs. We predict the signature partner band ($\alpha = -1/2$) of the $1/2^-$[521]($\alpha = +1/2$) band in ^{251}Md. A band crossing between the $1/2$[521]($\alpha = -1/2$) and $1/2$[770]1/2($\alpha = -1/2$) configurations occurs at very low frequency (Fig. 1). Therefore, our calculation predict a sharp back-bending of the $\alpha = -1/2$ band would take place at $\hbar\omega \approx 0.15$ MeV.

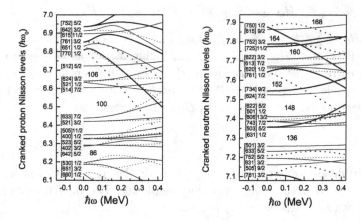

Fig. 1. The cranked Nilsson orbitals near the Fermi surface of ^{253}No. The signature $\alpha = +1/2$ and $\alpha = -1/2$ orbitals are denoted by solid and dotted lines, respectively. The high-j intruder orbitals are denoted by bold lines. $\varepsilon_2 = 0.29, \varepsilon_4 = 0.02$.

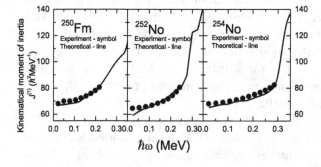

Fig. 2. Kinematic moment of inertia $\Im^{(1)}$ for the bands of ^{250}Fm and ^{252,254}No.

4. Summary

Rotational bands of ^{250}Fm, ^{252,253,254}No and ^{251}Md are studied by the PNC-CSM method. The observed experimental data are reproduced very well by our calculation. Backbendings are predicted at about $\hbar\omega \approx 0.275$ MeV for the bands of ^{252,253,254}No. This is mainly due to the rapidly aligned angular momentum of proton [770]1/2 pairs and neutron [761]1/2,

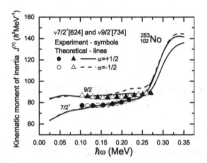

Fig. 3. Same as Fig.2, but for $\nu 7/2^+[624]$ and $\nu 9/2^+[734]$ bands of ^{253}No.

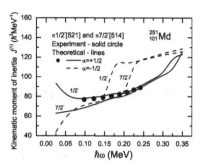

Fig. 4. Same as Fig.2, but for the $\pi 1/2^-[521]$ and $\pi 7/2^-[514]$ bands of ^{251}Md.

[734]9/2 pairs. We predict the signature partner band ($\alpha = -1/2$) of the $1/2^-[521](\alpha = +1/2)$ band in ^{251}Md. A sharp backbending of the $\alpha = -1/2$ band would take place at very low frequency, which due to the band crossing between the $1/2[521](\alpha = -1/2)$ and $1/2[770]1/2(\alpha = -1/2)$ configurations.

References

1. S. Hofmann and G. Münzenberg, *Rev. Mod. Phys.* 72, 733 (2000).
2. Yu.Ts. Oganessian, *J. Phys. G: Nucl. Part. Phys.* 34, R165 (2007).
3. Yu.Ts. Oganessian, *et al.*, *Phys. Rev. Lett.* 104, 142502 (2010).
4. R.-D. Herzberg and P.T. Greenlees, *Prog. Part. Nucl. Phys.* 61 (2008) 674.
5. P. Reiter *et al.*, *Phys. Rev. Lett.* 82, 509 (1999).
6. M. Leino *et al.*, *Eur. Phys. J. A* 6, 63 (1999).
7. P.A. Butler *et al.*, *Phys. Rev. Lett.* 89, 202501 (2002).
8. S. Eeckhaudt *et al.*, *Eur. Phys. J. A* 26, 227(2005).
9. R.-D. Herzberg *et al.*, *Phys. Rev. C* 65, 014303 (2001).
10. J.E. Bastin *et al.*, *Phys. Rev. C* 73, 024308 (2006).
11. P. Reiter *et al.*, *Phys. Rev. Lett.* 95, 032501 (2005).
12. A. Chatillon *et al.*, *Phys. Rev. Lett.* 98, 132503 (2007).
13. S. Ketelhut *et al.*, *Phys. Rev. Lett.* 102, 212501 (2007).
14. S. K. Tandel *et al.*, *Phys. Rev. Lett.* 97, 082502 (2006)
15. B. Sulignano *et al.*, *Eur. Phys. J. A* 33, 327 (2007).
16. P. T. Greenlees *et al.*, *Phys. Rev. C* 78, 021303(R) (2008).
17. H. B. Jeppesen *et al.*, *Phys. Rev. C* 79, 031303(R) (2009).
18. J. Qian *et al.*, *Phys. Rev. C* 79, 064319 (2009).
19. H. B. Jeppesen *et al.*, *Phys. Rev. C* 80, 034324 (2009).
20. Xiao-tao He *et al.*, *Nucl. Phys. A* 817, 45 (2009).
21. Zhen-hua Zhang *et al.*, *Phys. Rev. C* 83, 011304(R) (2011).
22. S.G. Nilsson *et al.*, *Nucl. Phys. A* 131, 1 (1969).

SHAPE COEXISTENCE AND SHAPE EVOLUTION IN ^{157}Yb[*]

H. HUA,[†] C. XU, X. Q. LI,[‡] J. MENG, F. R. XU, Y. SHI, S.Q. ZHANG, Z. Y. LI

School of Physics and State Key Laboratory of Nuclear Physics and Technology,
Peking University, Beijing 100871, China
[†]*E-mail: Hhua@hep.pku.edu.cn;* [‡]*E-mail: Llixq2002@hep.pku.edu.cn*

L. H. ZHU, X. G. WU,[§] G. S. LI, C. Y. HE

Department of Nuclear Physics, China Institute of Atomic Energy,
Beijing, 102413, China
[§]*E-mail: Wxg@ciae.ac.cn*

S. G. ZHOU

Institute of Theoretical Physics, Chinese Academy of Sciences, Beijing 100190, China;
Center of Theoretical Nuclear Physics, National Laboratory of Heavy Ion Accelerator,
Lanzhou 730000, China
E-mail: Sgzhou@itp.ac.cn

S. Y. Wang

Department of Space Science and Applied Physics,
Shandong University at Weihai, Weihai 264209, China
E-mail: Sywang@sdu.edu.cn

High-spin states in ^{157}Yb have been populated in the ^{144}Sm(^{16}O,3n)^{157}Yb fusion-evaporation reaction at a beam energy of 85 MeV. Two rotational bands built on the $\nu f_{7/2}$ and $\nu h_{9/2}$ intrinsic states, respectively, have been established for the first time. The structural characters observed in ^{157}Yb provide evidence for shape coexistence of three distinct shapes: prolate, triaxial, and oblate.

[*]This work is supported by the Major State Basic Research Development Program (2007CB815002) and by the National Natural Science Foundation of China (10775005, 10875002, 10735010, 10975007, 10875157, and J0730316) etc.

1. Introduction

The light rare-earth nuclei around A = 150 mass region, which locate above the double-closed shells N = 82 and Z = 64, have attracted a lot of experimental and theoretical studies. For example, these experimental studies led to the identification of many interesting phenomena: the pronounced vibrational nature of the low-lyingmembers of the ground state band; the occurrence of an island of long-lived isomers; the superdeformed nuclear shape with a major-to-minor axis ratio of approximately 2; the shape transition from prolate collective to oblate noncollective rotation via the mechanism of band termination between spins I ~ 30–42. At the same time, theoretical studies of the nuclear structure properties over a wide range of isospin, not only for the ground states but also for the low and high-spin states, have been made to try to understand systematically themicroscopic origin of these rich phenomena.

Due to the competition of the collective vibrational, rotational, and single-particle degrees of freedom, the level schemes in these nuclei are very complex even in the low-excitation-energy region. Thus, the spectroscopy of odd-*A* nuclei is particularly useful in identifying the active single-particle orbitals and providing more unambiguous clues for understanding the mechanism behind the shape evolution and shape coexistence in the transitional region. Here, we will report our recent work on the structure of ^{157}Yb.

2. Experiment and results

The present experiment was performed at the HI-13 tandem facility of the China Institute of Atomic Energy. The ^{144}Sm(^{16}O, 3n)^{157}Yb reaction was used to populate the high-spin states in ^{157}Yb. The target was ^{144}Sm with a thickness of 3.1mg/cm^2. It is on a 13.2 mg/cm^2 Pb backing. The deexcitation γ rays were detected by a γ detector array that consists of 10 Compton-suppressed HPGe detectors and one clover HPGe detector. Two additional planar HPGe detectors were also used to detect low-energy γ rays. The beam energies of 85 MeV were chosen to produce the ^{157}Yb. Coincidence data were recorded when two or more HPGe detectors were fired. A symmetric γ -γ matrix and a two dimensional angular correlation matrix were built. The level scheme analysis was performed using the RADWARE program [1].

The partial level scheme of ^{157}Yb, deduced from the present work, is shown in Fig. 1. It was constructed from γ-γ coincidence relationships, intensity balances, and DCO analyses. The present analysis confirms most of the transitions found in the previous study [2], and only a few level sequences have been reordered. The $13/2^+$ isomer has been reported to have an excitation energy of 528.8 keV in ^{157}Yb [2,3] and decays to the $\nu f_{7/2}$ ground state via two parallel transition ascades. In the present work, a coincidence spectrum gated with the γ-ray 205.5-keV transition shows the presence of a new 288.9-keV line, which was also found to be in coincidence with the above transition cascades but not with the 494.4-keV transition. With the new observation of 288.9-keV transition and their coincidence relationships, the order of the 323.3-and 205.5-keV transitions is fixed. The DCO ratio analyses and intensity considerations suggest that both the 205.5- and 288.9-keV transitions are likely to have the same multipolarities (mixed $M1/E2$). The spin and parity $9/2^-$ are tentatively assigned to the level at 205.5 keV. Two collective bands built on the $\nu f_{7/2}$ and $\nu h_{9/2}$ intrinsic states, respectively, were observed for the first time.

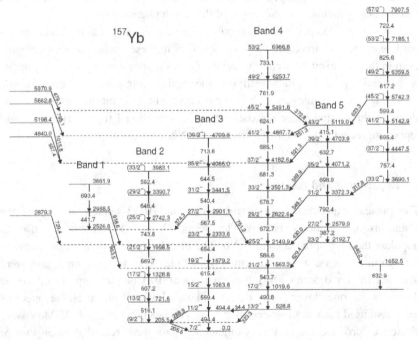

Fig. 1. Partial level scheme of ^{157}Yb. Energies are in keV.

3. Discussion

In order to have a further understanding of the structural properties of ^{157}Yb, Cranked Woods-Saxon-Strutinsky calculations have been performed by means of Total-Routhian-Surface (TRS) methods in a three-dimensional deformation space (β_2, , β_4) [4, 5]. The calculated equilibrium deformations for the band-head of band 2, band 3 and band 4 are listed in Table 1.

The irregular positive-parity and irregular negative-parity sequence in [155]Er have been suggested to have the structures dominated by the oblate configuration of $\nu(f_{7/2}^3)(h_{9/2})(i_{13/2})$ and $\nu(f_{7/2}^2)(h_{9/2})(i_{13/2}^2)$, respectively. These two sequences in [155]Er are very similar to the sequences of band 5 and band 6 in [157]Yb. This striking similarity and the irregular $E2$ transition sequences of the band 5 and band 6 in [157]Yb indicate that the band 5 and band 6 have most likely the same oblate configurations, respectively. Thus, based on the theoretical calculation and systematic comparison, the prolate, triaxial and oblate shapes were found to coexist in [157]Yb.

Table 1. The calculated equilibrium deformations for the band-head of band 2, band 3 and band 4

Band	Band-head	β_2	β_4	
2	$9/2^-$	0.157	0.024	~ 3
3	$7/2^-$	0.160	0.032	~ 0
4	$13/2^+$	0.165	0.044	~ 21

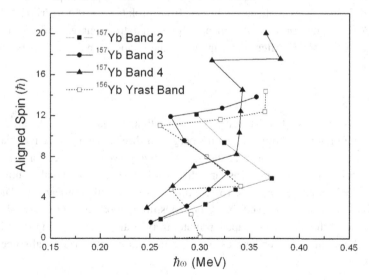

Fig. 2. Aligned spins as a function of the rotational frequency for the band 2, 3, 4 in [157]Yb and yrast band in [156]Yb.

These bands have also been studied using the particle-rotor model [6, 7]. The deformation parameters used are taken from TRS results. An off-diagonal Coriolis attenuation parameter is introduced and a variable moment of inertia is also used. The calculated energy spectra agree quite well with experimental data. According to the calculation, the band 2 and band 3 are built mainly on the single-particle configuration of $1/2^-[541]$(from $\nu f_{7/2}$) and $1/2^-[530]$ (from $\nu h_{9/2}$), respectively, though there is a strong mixing between these two configurations.

Experimental alignments as a function of rotational frequency for bands 2, 3, and 4 in ^{157}Yb, together with that for the yrast band in ^{156}Yb, are plotted in Fig. 2. In Fig. 2, bands 2 and 3 show a similar sharp backbend around rotational frequencies of 0.30–0.33 MeV. In Ref. [8], the first band crossing occurring in the neighboring ^{156}Yb nucleus was ascribed to the alignment of a pair of $h_{11/2}$ protons. The similar crossing frequencies between bands 2 and 3 of ^{157}Yb and the yrast band of ^{156}Yb indicate that the $h_{11/2}$ proton-pair alignment may also be responsible for the band crossings observed in bands 2 and 3 of ^{157}Yb. The calculations show that the first band crossing in the $\nu f_{7/2}$ band of ^{157}Yb caused by an $h_{11/2}$ proton-pair alignment occurs around a rotational frequency of 0.30 MeV, which reproduces well the experimental value. The small variation in the crossing frequencies for the $h_{11/2}$ proton-pair alignments observed in bands 2 and 3 of ^{157}Yb may be due to the small variation in their deformation.

According to the TRS calculations, with the first $h_{11/2}$ proton-pair alignment, the band 3 undergoes a shape evolution: from prolate ($\gamma \approx 0°$) to triaxial ($\gamma \approx -39°$). An $i_{13/2}$ neutron-pair alignment is predicted to occur around rotational frequency 0.38 MeV. After this alignment, the shape of band 3 will become oblate.

5. Summary

The level structures of ^{157}Yb have been investigated via ^{144}Sm(^{16}O, 3n)^{157}Yb fusion-evaporation reaction at beam energy 85 MeV. Two collective bands built on the $\nu f_{7/2}$ and $\nu h_{9/2}$ intrinsic states, respectively, were observed for the first time. Experimental observations are discussed in terms of Total-Routhian-Surface (TRS) methods, particle-rotor model and systematic comparison. These structural characters observed in ^{157}Yb constitute good evidences for shape coexistence of three distinct shapes: prolate, triaxial and oblate. At higher spins, the $\nu f_{7/2}$ band in ^{157}Yb undergoes a shape evolution with sizable alignments occurring.

Acknowledgments

The authors wish to thank Dr. Q. W. Fan for making the target and the staff in the tandem accelerator laboratory at the China Institute of Atomic Energy (CIAE), Beijing.

References

1. D.C. Radford, *Nucl. Instrum. Methods Phys. Res.* **A361**, 297(1995).
2. Y. Zheng et al., *Eur. Phys. J.* **A14**, 133(2002).
3. M.H. Rafailovich, O.C. Kistner, A.W. Sunyar, S. Vajda, and G.D. Sprouse, *Phys. Rev.* **C30**, 169(1984).
4. W. Satula, R. Wyss, and P. Magierski, *Nucl. Phys.* **A578**, 45(1994).
5. F.R. Xu, W. Satula, and R. Wyss, *Nucl. Phys.* **A669**, 119(2000).
6. S. Q. Zhang, B. Qi, S. Y. Wang, J. Meng, *Phys. Rev.* **C75**, 044307(2007).
7. B. Qi, S. Q. Zhang, S. Y. Wang, J. Meng, *Int. Jour. Mod. Phys.* **E18**, 109(2009).
8. Z.Y. Li, H. Hua et al., *Phys. Rev.* **C77**, 064323(2008).

MULTISTEP SHELL MODEL METHOD IN THE COMPLEX ENERGY PLANE

R. J. LIOTTA

KTH, Alba Nova University Center, SE-10691 Stockholm, Sweden

We have adopted the multistep shell model in the complex energy plane to study excitations occurring in the continuum part of the nuclear spectrum. In this method of solving the shell model equations one proceeds in several steps. In each step one constructs building blocks to be used in future steps. We applied this formalism to analyze the unbound nucleus ^{12}Li. In this case the excitations correspond to the motion of three particles partitioned as the product of a one-particle and two-particle systems.

Keywords: Shell mode; continuum; drip line; complex energy plane.

1. Introduction

The study of unstable nuclei is a very difficult undertaking. However, the system may be considered stationary if it lives a long time. In this case the time dependence can be circumvented. On the other hand, experimental facilities allow one nowadays to measure systems living a very short time. To describe these short time processes one has to consider the decaying character of the system. Of the various theories that have been conceived to analyze unbound systems, we mention an extension of the shell model to the complex energy plane.[1] The basic assumption of this theory is that resonances can be described in terms of states lying in the complex energy plane. The real parts of the corresponding energies are the positions of the resonances while the imaginary parts are minus twice the corresponding widths, as it was proposed by Gamow at the beginning of quantum mechanics.[2] These complex states correspond to solutions of the Schrödinger equation with outgoing boundary conditions. Detials of the formalism and its application to two-nucleon systems can be found, e.g., in Refs.[3-5]

Our aim is to develop a suitable formalism to treat unstable nuclei involving many valence nucleons in the continuum.[6] This formalism is an extension of the shell model in the complex energy plane.[1] The correlations induced by the pairing force acting upon particles moving in decaying single-particle states is taken into account by using the multistep shell model (MSM).[7] The formalism is presented in Section 2. Applications are in Section 3 and a summary and conclusions are in Section 4.

2. The formalism

The eigenstates of a central potential obtained as outgoing solutions of the Schrödinger equation can be used to express the Dirac δ-function as[8]

$$\delta(r - r') = \sum_n w_n(r)w_n(r') + \int_{L^+} dE u(r, E)u(r', E), \tag{1}$$

where the sum runs over all the bound and antibound states plus the complex states (resonances) which lie between the real energy axis and the integration contour L^+. The antibound states are virtual states with negative scattering length. The wave function of a state n in these discrete set is $w_n(r)$ and $u(r, E)$ is the scattering function at energy E.

Discretizing the integral of Eq. (1) one obtains the set of orthonormal vectors $|\varphi_j\rangle$ forming the Berggren representation.[9] Since this discretization provides an approximate value of the integral, the Berggren vectors fulfill the relation $I \approx \sum_j |\varphi_j\rangle\langle\varphi_j|$, where all states, that is bound, antibound, resonances and discretized scattering states, are included. Using the Berggren representation one readily gets the two-particle shell-model equations in the complex energy plane (CXSM),[4] i.e.,

$$(W(\alpha_2) - \epsilon_i - \epsilon_j)X(ij; \alpha_2) = \sum_{k \leq l}\langle \tilde{kl}; \alpha_2|V|ij; \alpha_2\rangle X(kl; \alpha_2), \tag{2}$$

where V is the residual interaction. The tilde in the interaction matrix element denotes mirror states so that in the corresponding radial integral there is not any complex conjugate, as required by the Berggren metric. The two-particle states are labeled by α_2 and Latin letters label single-particle states. $W(\alpha_2)$ is the correlated two-particle energy and ϵ_i is single-particle energy. The two-particle wave function is given by

$$|\alpha_2\rangle = P^+(\alpha_2)|0\rangle, \tag{3}$$

where the two-particle creation operator is given by

$$P^+(\alpha_2) = \sum_{i \leq j} X(ij; \alpha_2)\frac{(c_i^+ c_j^+)_{\lambda_{\alpha_2}}}{\sqrt{1 + \delta_{ij}}}, \tag{4}$$

and λ_{α_2} is the angular momentum of the two-particle state.

2.1. *The Multistep Shell Model Method*

The Multistep Shell Model Method (MSM) solves the shell model equations in several steps. In the first step the single-particle representation is chosen. In the second step the energies and wave functions of the two-particle system are evaluated with a given two-particle interaction. The three-particle states are evaluated in terms of a basis consisting of the tensorial product of the one- and two-particle states previously obtained. In this and subsequent steps the interaction does not appear explicitly in the formalism. Instead, it is the wave functions and energies of the components of the MSM basis that replace the interaction. The MSM basis is overcomplete and nonorthogonal. To correct this one needs to evaluate the overlap matrix among the basis states also. A general description of the formalism is in Ref.[10] The particular system that is of our interest here, i.e., the three-particle case, can be found in Ref.[7] Below we refer to this formalism as CXMSM.

The three-particle energies $W(\alpha_3)$ are given by[7]

$$(W(\alpha_3) - \varepsilon_i - W(\alpha_2))\langle \alpha_3 | (c_i^+ P^+(\alpha_2))_{\alpha_3} | 0 \rangle$$

$$= \sum_{j\beta_2} \left\{ \sum_k (W(\beta_2) - \varepsilon_i - \varepsilon_k) A(i\alpha_2, j\beta_2; k) \right\} \langle \alpha_3 | (c_j^+ P^+(\beta_2))_{\alpha_3} | 0 \rangle, \quad (5)$$

where

$$A(i\alpha_2, j\beta_2; k) = \hat{\alpha}_2 \hat{\beta}_2 \begin{Bmatrix} i & k & \beta_2 \\ j & \alpha_3 & \alpha_2 \end{Bmatrix} Y(kj; \alpha_2) Y(ki; \beta_2), \quad (6)$$

$$Y(ij; \alpha_2) = (1 + \delta(i, j))^{1/2} X(ij; \alpha_2), \quad (7)$$

and the rest of the notation is standard.

The matrix defined in Eq. (5) is not hermitian and the dimension may be larger than the corresponding shell-model dimension. This is due to the violations of the Pauli principle as well as overcounting of states in the CXMSM basis. Therefore the direct diagonalization of Eq. (5) is not convenient. One needs to calculate the overlap matrix in order to transform the CXMSM basis into an orthonormal set. In this three-particle case the overlap matrix is

$$\langle 0 | (c_i^+ P^+(\alpha_2))_{\alpha_3}^\dagger (c_j^+ P^+(\beta_2))_{\alpha_3} | 0 \rangle = \delta_{ij} \delta_{\alpha_2 \beta_2} + \sum_k A(i\alpha_2, j\beta_2; k). \quad (8)$$

Using this matrix (8) one can transform the matrix determined by Eq. (5) into a hermitian matrix T which has the right dimension. The diagonalization of T provides the three-particle energies. The corresponding wave function amplitudes can be readily evaluated to obtain

$$|\alpha_3\rangle = P^+(\alpha_3)|0\rangle, \tag{9}$$

$$P^+(\alpha_3) = \sum_{i\alpha_2} X(i\alpha_2; \alpha_3)(c_i^+ P^+(\alpha_2))_{\alpha_3}, \tag{10}$$

where $P^+(\alpha_3)$ is the three-particle creation operator. It has to be pointed out that in cases where the basis is overcomplete the amplitudes X are not well defined. But this is no hinder to evaluate the physical quantities. For details see Ref.[7]

The CXMSM allows one to choose in the basis states a limited number of excitations. This is because in the continuum the vast majority of basis states consists of scattering functions. These do not affect greatly physically meaningful two-particle states. That is, the majority of the two-particle states provided by the CXSM are complex states which form a part of the continuum background. Only a few of those calculated states correspond to physically meaningful resonances, i.e., resonances which can be observed. Below we call a "resonance" only to a complex state which is meaningful. These resonances are mainly built upon single-particle states which are either bound or narrow resonances. Yet, one cannot ignore the continuum when evaluating the resonances. The continuum configurations in the resonance wave function are small but many, and they affect the two-particle resonance significantly.[4] Therefore, the great advantage of the CXMSM is that one can include in the basis only two-particle resonances, while neglecting the background continuum states, which form the vast majority of complex two-particle states.

3. Applications

Using the Berggren single-particle representation described above, we will evaluate the complex energies and wave functions of the unbound nucleus ^{12}Li using the MSM basis states consisting of the Berggren one-particular states, which are states in ^{10}Li, times the two-particle states corresponding to ^{11}Li. The spectrum of ^{11}Li was already evaluated within the CXSM including antibound states.[13] Here we will repeat that calculation in order to determine the two-particle states to be used in the calculation of the three-particle system, i.e., ^{12}Li.

To define the Berggren single-particle representation we still have to choose the integration contour L^+ (see Eq. (1)). The valence shells are the low lying resonances $0p_{1/2}$ at (0.195,-0.047) MeV and $0d_{5/2}$ at (2.731,

-0.545) MeV. Besides, the state $1s_{1/2}$ appears as an antibound state. To include in the representation the antibound $1s_{1/2}$ state as well as the Gamow resonances $0p_{1/2}$ and $0d_{5/2}$ we will use two different contours. The number of points on each contour defines the energies of the scattering functions in the Berggren representation, i.e., the number of basis states corresponding to the continuum background. This number is not uniformity distributed, since in segments of the contour which are close to the antibound state or to a resonance the scattering functions increase strongly.

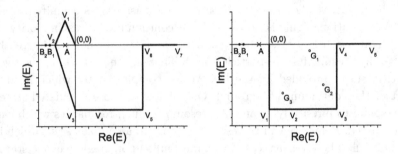

Fig. 1.   Left: Contour used to include the antibound state (see, also, Ref.[12]). The points B_i denote bound state energies while A denotes the antiboud state. Right: Contour used to include the Gamow resonances represented by the points G_i.

The dynamics of ^{11}Li is determined by the pairing force acting upon the two neutrons coupled to a state 0^+, which behaves as a normal even-even ground state.[11,13] Besides the energy, this state has been measured to have an angular momentum contain of about 60 % of s-waves and 40 % of p-waves, although small components of other angular momenta are not excluded.

We will use a separable interaction as in Ref.[13] The strength G_{λ_2}, corresponding to the states with angular momentum λ_2 and parity $(-1)^{\lambda_2}$, will be determined by fitting the experimental energy of the lowest of these states, as usual. It is worthwhile to point out that G_{λ_2} defines the Hamiltonian and, therefore, is a real quantity. The energies are thus obtained by solving the corresponding dispersion relation. The two-particle wave function amplitudes are given by

$$X(ij; \alpha_2) = N_{\alpha_2} \frac{f(ij, \alpha_2)}{\omega_{\alpha_2} - (\epsilon_i + \epsilon_j)},$$
(11)

where $f(ij, \alpha_2)$ is the single particle matrix element of the field defining the separable interaction and N_{α_2} is the normalization constant determined by the condition $\sum_{i \leq j} X(ij; \alpha_2)^2 = 1$.

Due to the large number of scattering states included in the single-particle representation the dimension of three-particle basis is also large. The scattering states are needed in order to describe these unstable states. In the calculations we took into account all the possibilities described above regarding the energies of the single-particle state $0p_{1/2}$ as well as the binding energy of the state ^{11}Li(gs).

With the single-particle states and the two-particle states ^{11}Li(gs) and ^{11}Li(2_1^+) discussed above, we formed all the possible three-particle basis states. We found that the only physically relevant states are those which are mainly determined by the bound state ^{11}Li(0_1^+). The corresponding spins and parities are $1/2^+$, $1/2^-$ and $5/2^+$. States like $3/2^+$, which arises from the CXMSM configuration $|1s_{1/2} \otimes 2_1^+; 3/2^+ >$, is not a meaningful state. The corresponding calculated energies are shown in Fig. 2.

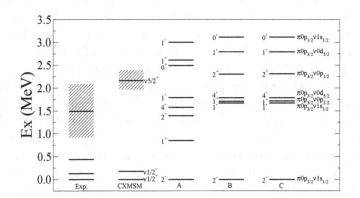

Fig. 2. Experimental level scheme in ^{12}Li. The three lowest levels are from,[16] while the one at 1.5 MeV is from.[15] In the second column are the three-neutron CXMSM results. In the columns A-C are the shell-model calculations corresponding to different truncation schemes: A) Maximum of 1p excitation from p to sd shell, B) maximum of 3p excitations and C) full psd space. Dashed lines indicate the widths of the resonances.

In Ref.[16] it was also found that ^{12}Li(gs) is an antibound state but, in addition, two other low-lying states were observed at 0.250 MeV and 0.555 MeV by using two-proton removal reactions. In this case the $0p_{3/2}$ proton in the core may interfere with the neutron excitations evaluated above. In particular, the antibound $1/2^+$ ground state would provide, through the proton excitation, a state 1^- and a 2^-. This is the situation encountered in the shell model calculation.[17] One sees in this Figure that the full calculation predicts all excited states to lie well above the corresponding experimental values. It is worthwhile to point out that the calculated states exhibit rather pure shell model configurations. For instance the states 2_1^- (ground state) and 1_1^- are mainly composed of the configuration $|\pi [0p_{3/2}] \nu [(0p_{1/2})^2 1s_{1/2}]\rangle$. This does not fully agree with our CXMSM calculation, since in our case this wave function is mainly of the form $|1s_{1/2} \otimes ^{11}\text{Li(gs)}\rangle$. This differs from the shell model case in two ways. First, the state $^{11}\text{Li(gs)}$ contains nearly as much of $1s_{1/2}$ as of $0p_{1/2}$. Second the continuum states contribute much in the building up of the antibound $^{12}\text{Li(gs)}$ wave function, as discussed above. In our representation it is straightforward to discern the antibound character of this state, which is not the case when using harmonic oscillator bases.

4. Summary and conclusions

We have studied excitations occurring in the continuum part of the nuclear spectrum which are at the limit of what can be observed within present experimental facilities. These states are very unstable but yet live a time long enough to be amenable to be treated within stationary formalisms. We have thus adopted the CXSM (shell model in the complex energy plane[4]) for this purpose. In addition we performed the shell model calculation by using the multistep shell model. In this method of solving the shell model equations one proceeds in several steps. In each step one constructs building blocks to be used in future steps.[10] We applied this formalism to analyze ^{12}Li as determined by the neutron degrees of freedom, i.e., the three protons in the core were considered to be frozen. In this case the excitations correspond to the motion of three particles, partitioned as the one-particle times the two-particle systems.

By using single-particle energies (i.e., states in ^{10}Li) as provided by experimental data when available or as provided by our calculation, we found that the only physically meaningful two-particle states are $^{11}\text{Li(gs)}$, which is a bound state, and $^{11}\text{Li}(2_1^+)$, which is a resonance. As a result there are only three physically meaningful states in ^{12}Li which, besides the antibound ground state, it is predicted that there is a resonance $1/2^-$ lying at about 1 MeV and about 800 keV wide and another resonance which is $5/2^+$ lying at about 1.1 MeV and 500 keV wide. That the ground state is an antibound (or virtual) state was confirmed by a number of experiments[14–16] and the state $5/2^+$ has probably been observed in.[15] However, in[16] two additional states, lying at rather low energies, have been observed which do not seem to correspond to the calculated levels. It has to be mentioned that neither a shell model calculation, performed within an harmonic oscillator basis, provides satisfactory results in this case. Yet, we found that this shell model calculation works better than one would assume given the unstable character of the states involved.

Acknowledgments

The author would like to thank Prof. F. Xu for inviting him to the conference and for fruitful discussions. The collaboration of Chong Qi is grateful acknowledged. This work has been supported by the Swedish Research Council under grant Nos. 623- 2009-7340 and 2010-4723.

References

1. R. Id Betan, R. J. Liotta, N. Sandulescu, and T. Vertse, *Phys. Rev. Lett.* **89**, 042501 (2002).
2. G. Gamow, *Z. Phys.* **51**, 204 (1928).
3. N. Michel, W. Nazarewicz, M. Ploszajczak, and K. Bennaceur, *Phys. Rev. Lett.* **89**, 042502 (2002).
4. R. Id Betan, R. J. Liotta, N. Sandulescu, and T. Vertse, *Phys. Rev. C* **67**, 014322 (2003).
5. N. Michel, W. Nazarewicz, M. Płoszajczak, and T. Vertse, *J. Phys. G* **36**, 013101 (2009).
6. Z. X. Xu, R. J. Liotta, C. Qi, T. Roger, P. Roussel-Chomaz, H. Savajols, R. Wyss, *Nucl. Phys. A* **850**, 53 (2011).
7. J. Blomqvist, R. J. Liotta, L. Rydstrom, and C. Pomar, *Nucl. Phys. A* **423**, 253 (1984).
8. T. Berggren, *Nucl. Phys. A* **109**, 265 (1968).
9. R. J. Liotta, E. Maglione, N. Sandulescu, and T. Vertse, *Phys. Lett. B* **367**, 1 (1996).
10. R. J. Liotta and C. Pomar, *Nucl. Phys. A* **382**, 1 (1982).
11. H. Esbensen, G.F. Bertsch, and K. Hencken, *Phys. Rev. C* **56**, 3054 (1997), and references therein.
12. R. Id Betan, R. J. Liotta, N. Sandulescu, and T. Vertse, *Phys. Lett. B* **584**, 48 (2004).
13. R. Id Betan, R. J. Liotta, N. Sandulescu, T. Vertse, and R. Wyss, *Phys. Rev. C* **72**, 054322 (2005).
14. Yu. Aksyutina *et al.*, *Phys. Lett. B* **666**, 430 (2008).
15. T. Roger, PhD thesis; P. Roussel Chomaz *et al.*, to be published.
16. C. C. Hall *et al.*, *Phys. Rev. C* **81**, 021302(R) (2010).
17. C. Qi and F. R. Xu, *Chin. Phys. C* **32** (S2), 112 (2008).

THE EVOLUTION OF PROTONEUTRON STARS WITH KAON CONDENSATE

ANG LI

Department of Physics and Institute of Theoretical Physics and Astrophysics,
Xiamen University, Xiamen 361005, P. R. China
E-mail: liang@xmu.edu.cn
www.xmu.edu.cn

Protoneutron star models are constructed using a realistic equation of state of hot dense matter, and under three different strongly idealized stages concerning stellar evolution. Stability of protoneutron stars with respect to kaon condensate is studied.

Keywords: Protoneutron stars; dense matter; kaon condensate.

1. Introduction

Protoneutron stars are newly born hot and lepton rich objects, quite different from ordinary low-temperature lepton-poor neutron stars. They transform into standard neutron stars on a timescale of the order of ten seconds, needed for the loss of a significant lepton number excess via emission of neutrinos trapped in dense hot interior. The scenario of transformation of a protoneutron star into a neutron star could be strongly influenced by a phase transition in the central region of the star, such as a kaon condensation which is very possible at supranuclear densities.

With extending our previous work[1] to hot matter, we intend investigating the impact of a kaon condensate on proto-neutron star (PNS) matter at finite temperature and on the final PNS observables, combining a microscopic Brueckner-Hartree-Fock (BHF) approach for the baryonic part of the matter with a standard chiral model for the kaon-nucleon contribution.

Our paper is organized as follows. In Sec. II we introduce the theoretical models. The numerical results about the structure of PNS's, are then illustrated in Sec. III, and conclusions are drawn in Sec. IV.

2. Theoretical models

We employ the BHF approach for asymmetric nuclear matter at finite temperature[2] to calculate the baryonic contribution to the EOS of stellar matter. As to kaon condensation, for the required extension to finite temperature we employ the formalism of Ref. [3], which treats fluctuations around the condensate within the framework of chiral symmetry. For small condensate amplitudes this approach is exactly equivalent to the meson-exchange mean-field models of Ref. [4].

The kaon-nucleon free energy density obtained in this way is:[2]

$$f_{KN} = \omega_{KN} + \mu_K q_K \tag{1}$$

$$= f^2 \left[m_K^{*2}(1 - \cos\theta) + \mu_K^2 \frac{\sin^2\theta}{2} \right] + \mu_K q_K^{th} + \omega_{KN}^{th}$$

In Eq. (1) we adopt the 'standard' KN interaction parameters $a_1 m_s = -67$ MeV, $a_2 m_s = 134$ MeV, and $a_3 m_s = -134, -222, -310$ MeV to perform our numerical calculations, where the different choices of a_3 correspond to different values of the strangeness content of the proton, $y = 2\langle p|\bar{s}s|p\rangle/\langle p|\bar{u}u + \bar{d}d|p\rangle \approx 0, 0.36, 0.5$, in the chiral model.

One can determine the ground state by minimizing the total grand-canonical potential density.[2] This minimization, together with the chemical equilibrium and charge neutrality conditions, leads to three coupled equations.[2] The critical density for kaon condensation is determined as the point above which a real solution with $\theta > 0$ for the coupled equations. Finally the stable configurations of a (P)NS can be obtained from the well-known hydrostatic equilibrium equations of Tolman, Oppenheimer, and Volkov.

3. Results

In the following we present the results of our numerical calculations regarding the structure of PNS's. Fig. 1 shows the corresponding mass – central density relations obtained with the micro three-body force (TBF) (upper panels) and the pheno TBF (lower panels) in the following cases: (a) no kaon condensate (left panels), (b) with kaon condensate using $a_3 m_s = -134$ MeV (middle panels) or $a_3 m_s = -222$ MeV (right panels). We consider three different strongly idealized stages of the PNS evolution: (i) $T = 30$ MeV, $Y_e = 0.4$ (black dashed lines), the initial hot and neutrino-trapped state; (ii) $T = 30$ MeV, $x_\nu = 0$ (red dotted lines), the intermediate phase lasting about a few seconds, when neutrinos have diffused out of the still hot environment; (iii) $T = 0$ MeV, $x_\nu = 0$ (green solid lines), the final state of a cold NS formed after a few tens of seconds.

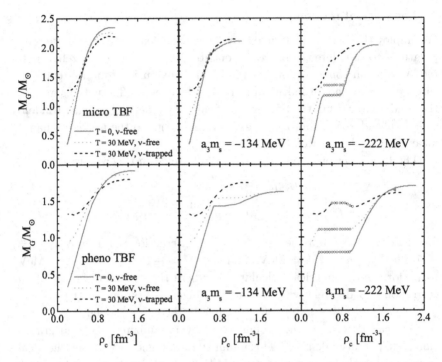

Fig. 1. (Color online) (Proto)neutron star mass – central density relations obtained with or without kaon condensation. The red triangle means the physically possible configuration ends here because a negative entropy appears (before the maximum mass is achieved).

In the case without kaons (left panels), the maximum mass of the PNS is slightly smaller than that of the NS, because neutrino trapping reduces the asymmetry of beta-stable matter. The presence of kaon condensation reverses the situation, and the PNS generally has a larger maximum mass than the NS. A delayed collapse scenario is therefore facilitated by the presence of a kaon condensate, as is indeed generally found.[2]

A rather extreme scenario is seen in the case of the pheno TBF, where a quite soft nuclear EOS (much softer than that of the micro TBF)[1] combined with a strong kaon condensate (lower right panel of Fig. 1), where actually the maximum mass of the NS remains higher than that of the PNS and no delayed collapse could occur. However, we consider this combination of extreme parameter choices unlikely, as it leads to unrealistically high central densities of the star. Furthermore, the case of a strong kaon condensate seems to be excluded in the present model now.[2]

4. Summary

In conclusion, we investigated the consequences as well as the EOS of including kaon condensation in hot and neutrino-trapped NS matter, under three different strongly idealized stages concerning PNS evolution. We found that kaons may lead a newly formed, hot PNS to metastability, that is, a delayed collapse while cooling down. It should be mentioned, however, that a delayed collapse scenario appear very probable in the form of either kaons or hyperons, that is, any type of strange hadrons.[5]

Acknowledgments

This work was funded by National Basic Research Program of China (Grant No 2009CB824800), National Natural Science Foundation of China (Grant No 10905048), and the Youth Innovation Foundation of Fujian Province (Grant No 2009J05013).

References

1. W. Zuo, A. Li, Z. H. Li, and U. Lombardo, *Phys. Rev. C* **70**, 055802 (2004), A. Li, G. F. Burgio, U. Lombardo, and W. Zuo, *Phys. Rev. C* **74**, 055801 (2006).
2. A. Li, X. R. Zhou, G. F. Burgio and H.-J. Schulze, *Phys. Rev. C* **81**, 025806 (2010) and references therein.
3. T. Tatsumi and M. Yasuhira, *Phys. Lett. B* **441**, 9 (1998).
4. J. A. Pons, J. A. Miralles, M. Prakash, and J. M. Lattimer, *Astrophys. J.* **553**, 382 (2001).
5. M. Prakash, I. Bombaci, M. Prakash, P. J. Ellis, J. M. Lattimer, and R. Knorren, *Phys. Rep.* **280**, 1 (1997).

HIGH SPIN STRUCTURES IN THE ^{159}Lu NUCLEUS

LI CONGBO

Department of Physics, Jilin University, Changchun 130023,
People's Republic of China
China Institute of Atomic Energy, PO Box 275, Beijing, People's Republic of China
E-mail: licb10@mails.jlu.edu.cn

LI XIANFENG, MA YINGJUN, ZHAO YANXIN, LIU GONGYE, LI LI

Department of Physics, Jilin University, Changchun 130023,
People's Republic of China

ZHU LIHUA, WU XIAOGUANG*, HE CHUANGYE, LI GUANGSHENG,
LIU YING, HAO XIN, LI XUEQIN

China Institute of Atomic Energy, PO Box 275, Beijing, Peoples Republic of China
*E-mail: wxg@ciae.ac.cn

High-spin states in ^{159}Lu were populated by fusion-evaporation reaction ^{144}Sm (^{19}F, 4n) at a beam of 106 MeV. The level scheme for ^{159}Lu is established up to spin $\frac{57}{2}^-$ with the addition of about 20 new transitions. The possible configurations of the updated bands are suggested by the experiments Routhians, alignments and $B(M1)/B(E2)$ ratios.

Keywords: High spin; quasiparticle configurations; $B(M1)/B(E2)$ ratios.

1. Introduction

The odd-Z, even-N isotopes in the rare earth and heavier-mass region can provide rich information on nuclear structure. There are several properties in the transitional nuclei with $64 \leq Z \leq 70$ and $82 \leq N \leq 90$. Small driving force can produce large changes in transitional region. The even-N Lu isotopes $^{161-167}$Lu have all been studied well. They all have been demonstrated by existence of wobbling excitation. At high spin, the even-even N=88 nuclei ^{154}Dy, ^{156}Er, ^{158}Yb, and odd-A nuclei ^{157}Tm, ^{157}Er, ^{157}Ho all have a shape from prolate collective to oblate non-collective via the mechanism of band termination. Drastic structure variation of a nucleus is expected when the nucleus approaches the transitional region where, with decreasing neutron number, the nucleus changes from a deformed to a spherical shape. Therefore it is interesting to study ^{159}Lu, which is very close to transitional region.

2. Experimental details and results

The high-spin states of ^{159}Lu were populated through the reaction ^{144}Sm (^{19}F, 4n) at a beam of 106 MeV ^{19}F ions provided by the HI-13 tandem accelerator of CIAE in Beijing. The target ^{144}Sm consisted of a self-supporting Sm foils of 1.2 mg/cm^2 thickness. Deexcitation γ-rays were recorded with an array of 12 HPGe detectors each surrounded by BGO anti-Compton shield and 2 planar-type HPGe detectors. A total of 2×10^8 events were recorded when two or more Ge detectors were in coincidence. In the offline analysis, the data were sorted into symmetry matrix and DCO matrix for the determination of γ-γ coincidence relationships and assigning levels spin.

On the basis of previous results [1], a new level scheme has been constructed on the basis of the γ-ray coincidence data and intensity balances. About 20 new transitions have been added into the level scheme, as shown in Fig. 1.

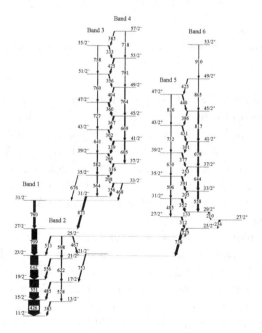

Fig. 1.   Proposed level scheme of ^{159}Lu from the present experiment.

3. Discussion

The structures of the bands in ^{159}Lu were interpreted in terms of quasiparticle configurations. The experimental Routhian e′ is plotted as a function of rotational frequency for the bands observed in ^{159}Lu in Fig. 2. The parameters J_0=12 MeV^{-1} $\hbar^2$ and J_1=104 MeV$^{-3}\hbar^4$ were obtained by reference to neighboring ^{158}Yb [2] and ^{160}Hf [3] even-even nuclei. In ^{159}Lu the band crossing occurs at a frequency of ω_c=0.33 and 0.32 MeV in the α=-1/2 and α=+1/2 negative parity yrast sequences. Compared with similar structures in ^{157}Tm and ^{155}Ho, it fits in well with the systematics of the first $i_{13/2}$ neutron backbend in the neutron-deficient rare earth nuclei.

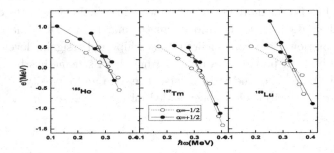

Fig. 2. Experimental routhians e′ as a function of ω for the yrast bands in ^{155}Ho, ^{157}Tm, ^{159}Lu.

The bands labeled 5 and 6 have the three quasiparticle configurations A_pAE, B_pAE respectively. A different three-quasi-particle structure with positive parity have been observed in the neighbors, ^{157}Ho [4], ^{157}Tm [5]. At ω=0.25 MeV bands 5 and 6 have aligments of 10.4 $\hbar$ and 10.0 $\hbar$, which fit well with CSM estimates. In Fig. 3, the aligned angular momentum values in bands 5 and 6 at low spin are lower in energy than bands 3 and 4, which is well consistent with the known behavior of AE and AB excitations in the N=88 even-even nuclei.

To further explore the possible configurations of ^{159}Lu, the experimental $B(M1)/B(E2)$ ratios were obtained, as showed in Fig. 4. The results of calculations for $B(M1)/B(E2)$ from the geometric model of Dönau show good agreement with the experimental data. It provides a further evidence for our proposed configuration assignments.

In summary, high spin states ($I \sim 30\hbar$) have been observed in the nucleus ^{159}Lu from γ-γ coincidence data. It has been possible to suggest

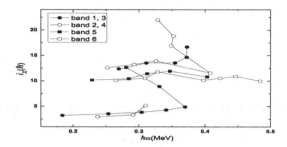

Fig. 3. The alignment i_x as a function of rotational frequency ω for all the bands in ^{159}Lu.

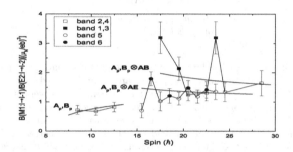

Fig. 4. Experimental $B(M1)/B(E2)$ ratios as a function of spin for bands in ^{159}Lu. The thick lines show results of calculations using the geometric model of Dönau and Frauendorf [6].

quasiparticle assignments for the observed bands. Routhians, alignments and $B(M1)/B(E2)$ ratios extracted from these data support the assignments for the observed bands.

References

1. Ma Yingjun et al., J. Phys. G: Nucl. Part. Phys. **21**, 937 (1995).
2. I. Ragnarsson al., Phys. Rev. Lett. **54**, 982 (1985).
3. M. Murzel et al., Nucl. Phys. A **516**, 189 (1990).
4. D.C. Radford et al., Nucl. Phys. A **545**, 1665 (1992).
5. M.A. Riley et al., Phys. Rev. C **51**, 1234 (1995).
6. F. Dönau, Nucl. Phys. A **471**, 469 (1987).

NUCLEAR STOPPING AND EQUATION OF STATE

QINGFENG LI

School of Science, Huzhou Teachers College,
Huzhou, 313000, P.R. China
E-mail: liqf@hutc.zj.cn

YING YUAN

Guangxi Teachers Education University,
Nanning, 530001, P.R. China

With the help of an updated transport model — ultra-relativistic quantum molecular dynamics, the nuclear stopping quantity *vartl* is used to investigate the stiffness of the equation of state of the nuclear matter as well as medium modifications of nucleon-nucleon elastic cross sections in heavy ion collisions (HICs) at SIS energies. It is found that the excitation function of the *vartl* value is sensitive to both the mean field and the two-body scattering. Hence, both parts should be taken into account consistently. At low SIS energies, in order to describe the FOPI data, the density dependence of the cross sections should be stronger implying that the effect of the nuclear structure may be noticed in the process of the HICs.

Keywords: Nuclear stopping; equation of state; transport model.

1. Introduction

The nuclear equation of state (EoS) is one of the key properties of the nuclear matter, which has being explored for several decades. It is well-known that it is important not only for the reaction but also for the structure related branches of the nuclear physics. Recently, one of the biggest progress in nuclear physics at low energies is the new announcement of the stiffness of the EoS of the nuclear matter.[1-3] Theoretically, an extended EoS should be further considered, i.e., the isospin dependent EoS (IEoS), in which the isospin asymmetric term is of importance to the experiments in which the isospin-asymmetric beams are used. The recent studies on the stiffness of the density dependent symmetry energy seem to almost pin it down at densities lower than the normal nuclear density.[2] However, a new puzzle occurs when to explore further the density dependence of the symmetry energy at supranormal densities.[4-6] Therefore, the stiffness of the symmetry energy at supranormal densities is still an open question and much more invetigations are needed.

The current status of the exploration of the EoS mentioned above does not indicate that the stiffness of the isoscalar part of the EoS has been undoubtedly confirmed since the medium modifications of the nucleon-nucleon cross sections are often neglected in the past investigations. Based on the theory of quantum hydrodynamics (QHD), both the mean field and the two-body collisions are from the same origin - the effective Lagrangian density, from which the medium effects on both the collision and the mean field should be considered self-consistently.[7-10]

In recent years, two experimental collaborations INDRA and FOPI of GSI at Darmstadt/Germany have published several systematic observations of a large amount of physical quantities such as the collective flows, the nuclear stopping, and the light cluster and new particle production.[11-13] After preliminary comparison with model simulations (mainly the quantum molecular dynamics (QMD) model), they found some discrepancies and it is hard to describe data only by varying the stiffness of the EoS.

Therefore, in this paper the nuclear stopping is investigated with an updated version of the ultra-relativistic quantum molecular dynamics (UrQMD) model, in which both various EoS and medium modifications of nucleon-nucleon elastic cross sections (NNECS) are considered at the same time. In the next section, the updated UrQMD model is briefly introduced. In section 3, results of the excitation function of the nuclear stopping for central Au+Au collisions at SIS energies are shown. Finally, a summary is given in section 4.

2. UrQMD model for HICs at SIS energies

The UrQMD model is based on analogous principles as the QMD model[14] and the relativistic quantum molecular dynamics (RQMD) model.[15] The first formal version (ver1.0) of the UrQMD transport model was published in the end of last century.[16] Since then, a large number of successful theoretical analyses, predictions, and comparisons with data based on this transport model have been accomplished for pp, pA and AA reactions for a large range of beam energies, i.e., from SIS, AGS, SPS, RHIC, up to LHC. At SIS energies, similar to QMD, hadrons are represented by Gaussian wave packets in phase space. The phase space of hadron i is propagated according to Hamilton's equation of motion, where the Hamiltonian H consists of the kinetic energy T and the effective two-body interaction potential energy V,

$$H = T + V, \tag{1}$$

and

$$T = \sum_i (E_i - m_i) = \sum_i (\sqrt{m_i^2 + \mathbf{p}_i^2} - m_i), \tag{2}$$

$$V = V_{Sky}^{(2)} + V_{Sky}^{(3)} + V_{Yuk} + V_{Cou} + V_{Pau} + V_{sym} + V_{mom}, \tag{3}$$

where the two-body and three-body Skyrme-, Yukawa-, Coulomb-, Pauli-, symmetry, and momentum dependent terms are included. In this work four parameter sets for the EoS are used for comparison: H-EoS, S-EoS, HM-EoS, and SM-EoS, which are described in Ref.[17]

In the previous work[18] the in-medium NNECS σ_{el}^* are treated to be factorized as the product of a medium correction factor $(F(u, \alpha, p))$ and the free NN elastic ones σ_{el}^{free}. For the inelastic channels σ_{in}, we still use the experimental free-space cross sections σ_{in}^{free}. It is believed that this assumption does not have strong influence on our present study at SIS energies. At present, three forms of in-medium NNECS are considered:

(1) σ^{free}, the free NNECS.

(2) σ_1^*, which is based on the extended QHD theory and reads as[18] $\sigma_1^* = F(u, \alpha, p)\sigma^{free}$ where the medium correction factor F depends on the nuclear reduced density $u = \rho_i/\rho_0$, the isospin-asymmetry $\alpha = (\rho_n - \rho_p)/\rho_i$ and the relative momentum of two colliding nuclei.

(3) σ_2^*, as in,[19] which reads as $\sigma_2^* = (1 - \xi u)\sigma^{free}$ where $\xi = 0.5$ for $E_{lab} < 0.25A$ GeV in this work.

Numerically, the *vartl* is defined as

$$vartl = \Gamma_{dN/dy_x}/\Gamma_{dN/dy_z}, \tag{4}$$

where Γ is the width of the y_x or y_z rapidity distribution of fragments, $\Gamma_{dN/dy_{x,z}} = \langle y_{x,z}^2 \rangle = \sum (y_{x,z}^2 N_{y_{x,z}})/N_{all}$. Here $N_{y_{x,z}}$ and N_{all} are yields of fragments in each x (or z) rapidity bin and in the whole rapidity range, respectively. It is easy to understand that *vartl* < 1 for an incomplete stopping or nuclear transparency while *vartl* > 1 means a strong transverse expansion or collectivity. Obviously, a full stopping occurs when *vartl* = 1.

3. *vartl* calculations

The excitation function of *vartl* for central Au+Au collisions from 0.09A GeV to 1.5A GeV is shown in Fig. 1 and also in Ref.[20] The FOPI data[21] is

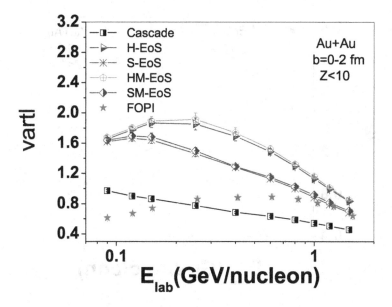

Fig. 1. The excitation function of *vartl* for central Au+Au collisions from 0.09A GeV to 1.5A GeV. The FOPI data is shown with stars while the UrQMD calculations with various EoS are shown with lines with symbols.

shown with stars while the UrQMD calculations with various EoS are shown with lines with symbols. The *vartl* value is calculated by using fragments with the proton number $Z < 10$. In calculations, the results with a cascade mode and with four EoS are compared with the data. The free NNECS are used. It is seen that the *vartl* value of the cascade mode is always less than 1 and decreases monotonously with increasing beam energy, which means less and less stopping. When the mean field is considered, the potentials reinforce the bound of nucleons and a stronger collectivity is shown in the transverse direction. In calculations with EoS, the softness of EoS gives a smaller *vartl* value while the momentum dependence term plays a negligible role. We also find that only a soft EoS can not describe the excitation function of FOPI data without considering any medium modifications on two-body collisions. Next, based on the result with the SM-EoS, we will further investigate the effect of medium modifications of NNECS on *vartl*.

Figure 2 illustrates the excitation function of *vartl* with the medium modified NNECS σ_1^* and the free one. It is seen clearly that a large reduction of cross sections at lower beam energies leads to large transparency so that the calculated *vartl* with σ_1^* is largely decreased at low SIS energies.

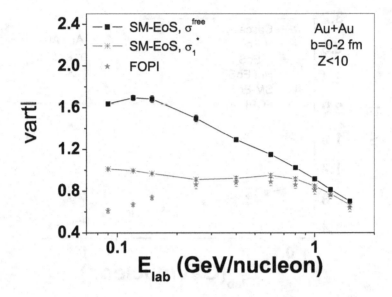

Fig. 2. The excitation function of *vartl* with the medium modified NNECS $\sigma_1{}^*$ as well as the free one.

At lower SIS energies (< 0.25A GeV), the result with $\sigma_1{}^*$ is still higher than data which implies that a stronger reduction factor on the NNECS might be required. It is proven by the calculation with $\sigma_2{}^*$, as done in Ref.[22]

4. Summary

We have presented the excitation function of the nuclear stopping described by *vartl* of light fragments for central Au + Au reactions with beam energies from 0.09A to 1.5A GeV. The UrQMD transport model has been used in all calculations. Based on the model we investigate the effects of both mean-field potentials and medium modifications of NNECS on nuclear stopping under the same initial and final freeze-out conditions. It is found that nuclear stopping is influenced by both the stiffness of the EoS and the medium modifications of NNECS for reactions at SIS energies.

The stronger reduction of cross sections seen in calculations at low SIS energies is interesting. The possible nuclear structure effect on the stability of emitted fragments might be important to the collective phenomena such as the stopping and the flows, which will be investigated in the near future.

Acknowledgments

We acknowledge support by the computing server C3S2 in Huzhou Teachers College. The work is supported in part by the key project of the Ministry of Education of China (No. 209053), the National Natural Science Foundation of China (Nos. 10905021,10979023), the Zhejiang Provincial Natural Science Foundation of China (No. Y6090210), and the Qian-Jiang Talents Project of Zhejiang Province (No. 2010R10102).

References

1. D. J. Magestro, W. Bauer and G. D. Westfall, *Phys. Rev.* **C62**, p. 041603 (2000).
2. B.-A. Li, L.-W. Chen and C. M. Ko, *Phys. Rept.* **464**, 113 (2008).
3. P. Danielewicz, R. Lacey and W. G. Lynch, *Science* **298**, 1592 (2002).
4. Z. Xiao, B.-A. Li, L.-W. Chen, G.-C. Yong and M. Zhang, *Phys. Rev. Lett.* **102**, p. 062502 (2009).
5. Z.-Q. Feng and G.-M. Jin, *Phys. Lett.* **B683**, 140 (2010).
6. W. Trautmann *et al.*, *Prog. Part. Nucl. Phys.* **62**, 425 (2009).
7. P. Danielewicz, *Annals Phys.* **152**, 239 (1984).
8. K. Chou, Z. Su, B. Hao and L. Yu, *Phys. Rept.* **118**, p. 1 (1985).
9. G. Mao, Z. Li, Y. Zhuo, Y. Han and Z. Yu, *Phys. Rev.* **C49**, 3137 (1994).
10. Q. Li, Z. Li and G. Mao, *Phys. Rev.* **C62**, p. 014606 (2000).
11. A. Andronic, J. Lukasik, W. Reisdorf and W. Trautmann, *Eur. Phys. J.* **A30**, 31 (2006).
12. W. Reisdorf *et al.*, *Nucl. Phys.* **A781**, 459 (2007).
13. W. Reisdorf *et al.*, *Nucl. Phys.* **A848**, 366 (2010).
14. J. Aichelin and H. Stoecker, *Phys. Lett.* **B176**, 14 (1986).
15. H. Sorge, H. Stoecker and W. Greiner, *Annals Phys.* **192**, 266 (1989).
16. M. Bleicher *et al.*, *J. Phys.* **G25**, 1859 (1999).
17. Q. Li, Z. Li, S. Soff, M. Bleicher and H. Stoecker, *J. Phys.* **G32**, 151 (2006).
18. Q. Li, Z. Li, S. Soff, M. Bleicher and H. Stoecker, *J. Phys.* **G32**, 407 (2006).
19. D. Klakow, G. Welke and W. Bauer, *Phys. Rev.* **C48**, 1982 (1993).
20. Y. Yuan, Q. Li, Z. Li and F.-H. Liu, *Phys. Rev.* **C81**, p. 034913 (2010).
21. W. Reisdorf *et al.*, *Phys. Rev. Lett.* **92**, p. 232301 (2004).
22. Y. Zhang, Z. Li and P. Danielewicz, *Phys. Rev.* **C75**, p. 034615 (2007).

COVARIANT DESCRIPTION OF THE LOW-LYING STATES IN NEUTRON-DEFICIENT Kr ISOTOPES

Z. X. LI, J. M. YAO*, Z. P. LI, J. XIANG and H. CHEN†

*School of Physical Science and Technology, Southwest University,
Chongqing, 400715, China*
*E-mail: jmyao@swu.edu.cn
†E-mail: chenh@swu.edu.cn

Starting from a covariant density functional of point-coupling, the spectroscopic properties of low-lying states in ^{72,74,76,78}Kr, including potential energy curves, energy spectrum and electric quadrupole transition strengths, are calculated by mixing of angular momentum projected axially deformed relativistic mean-field+BCS states. The calculated results are compared with the corresponding data. The phenomena of shape coexistence in these nuclei is discussed.

Keywords: Covariant energy functional; angular momentum projection; generator coordinate method; Krypton isotopes; low-lying states.

Nuclear shape is governed by a delicate interplay of the macroscopic liquid-drop properties of nuclear matter and microscopic shell effects and is therefore regarded as a very sensitive probe of the underlying nuclear structure. An atomic nucleus with different configurations can behave as different shapes, however, most of these shapes are unstable under the quantum fluctuation. Consequently, an unique shape is relatively stable and corresponds mostly the nuclear ground state. Previous studies already show that nuclei in some mass regions indeed have the phenomena of shape coexistence.[1] In these nuclei, there are at least two clearly distinguishable shapes, e.g., spherical versus deformed or prolate versus oblate. In addition, the energies of these shapes are nearly degenerate and the mixings of their configurations are very weak, i.e., negligible tunneling between different shapes. These characters become the necessary fingerprints in searching for the phenomena of shape coexistence in a single nucleus. Therefore, not only the ground state but also the low-energy/spin excited states are important in the studies of shape coexistence.

As one of the best examples for studying the phenomena of shape coexistence, the neutron-deficient Kr isotopes with a large quadrupole deformation have been extensively studied both experimentally and theoretically in recent years. On the experimental side, there are accumulated spectroscopic data for the low-lying states of Kr isotopes,[2-6] which provides rich information about the shape coexistence. In particular, recent results from low-energy Coulomb excitation of ^{74}Kr and ^{76}Kr provide the sign of the electric quadrupole moments for several low-lying states in these nuclei, proving the prolate character of the states in the ground-state band and oblate shapes for an excited configuration.[7] On the theoretical side, the first investigation of neutron-deficient Kr isotopes was performed for ^{72}Kr and a few neighboring odd nuclei using the Nilsson-Strutinsky approach.[8] Later on, the improved microscopic-macroscopic model,[9] the self-consistent mean-field models with the non-relativistic Skyrme[10] and the Gogny[11] forces as well as the relativistic Lagrangians,[12] have been used to calculate the potential energy curves, where both the oblate and prolate minima are observed in some light Kr isotopes.

In order to understand the phenomena of shape coexistence completely, one has to calculate the spectroscopic data of low-lying states, such as the excitation energies and electromagnetic transition matrix elements, which is beyond the mean-field level. Along this direction, the low-lying states of the even-even neutron-deficient $^{72-78}$Kr isotopes have been studied by mixing of axial mean-field configurations projected on both particle number and angular momentum based on the Skyrme force.[13] Although many of the global features are reproduced, there remain some evident discrepancies with experiment. Oblate ground-state shapes coexisting with prolate excited configurations were observed for all light Kr isotopes. The low-lying states in ^{72,74,76}Kr have been studied as well with the collective Bohr hamiltonian derived from the Hartree-Fock-Bogolyubov calculations with the Gogny D1S force including triaxial quadrupole deformations.[14] Good agreement was found with experimental excitation energies, transition probabilities, quadrupole moments, and charge radii. The structure of the low-lying states was found to be dominated by the coexistence of prolate and oblate shapes. The non-axial shapes were found to be essential to correctly describe the shape coexistence and shape transitions in the light Kr isotopes.

Recently, the framework of covariant density functionals (CDF) has been extended to include long-range correlations related to restoration of symmetries broken by the static mean field via angular-momentum projection

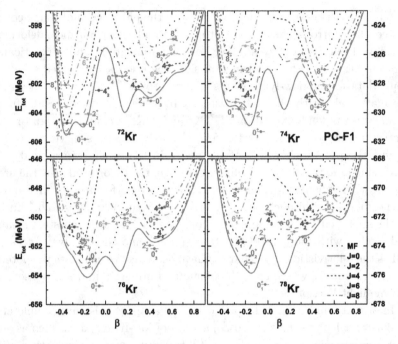

Fig. 1.   Mean-field and projected deformation energy curves for ^{72,74,76,78}Kr.

(AMP)[15] and to fluctuations of collective coordinates around the mean-field minimum via generator coordinate method (GCM).[16] This model has been used to perform detailed studies of low-energy collective excitation spectra and corresponding electromagnetic transition rates in Mg isotopes.[17] In this work, we apply this model to study the spectroscopic properties of low-lying states in ^{72,74,76,78}Kr. As the first step and to make comparison with the previous studies in Ref. 13, axial symmetry is restricted. The single-nucleon spinors are expanded in the basis of eigenfunctions of a three-dimensional harmonic oscillator in Cartesian coordinates, with 10 major shells. The PC-F1 set[18] is adopted. The details about the model calculations can be found in Refs. 15 and 16.

Figures 1 shows the mean-field and angular momentum projected energy curves for ^{72,74,76,78}Kr. The energies gained from the restoration of rotational symmetry are significant and even modify the topology of potential energy surfaces. It is seen that there are several distinguished minimal (prolate and oblate shapes in ^{72,78}Kr, prolate, spherical and oblate shapes in ^{74,76}Kr) in the projected 0^+ energy surfaces of these four light Kr isotopes, which implies the possible occurrence of shape coexistence.

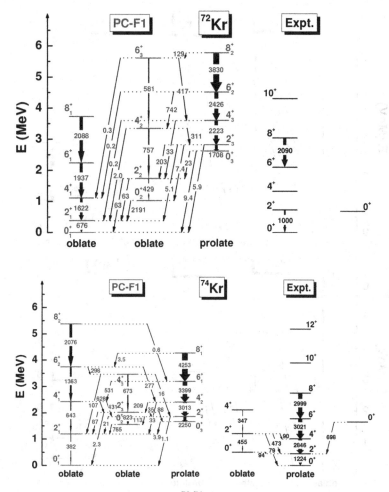

Fig. 2. Spectra of collective states for ^{72,74}Kr from present calculations (left) in comparison with the corresponding data (right). The arrows with numbers denote the $B(E2)$ transition strengths given in e^2 fm^4. Experimental data are taken from [7,19–21]. The labels "oblate" and "prolate" given to the bands correspond to the main components of the collective wave functions.

Figure 2 and 3 present the energy spectra of low-spin collective states for ^{72,74,76,78}Kr. The experiment data are given for comparison. The arrows with numbers denote the $B(E2)$ transition strengths given in e^2 fm ^{4}. The labels "oblate" or "prolate" indicate the main character of the collective wave functions. It is shown that the spectra can be reproduced qualitatively. The phenomena of shape coexistence is illustrated. Unfortunately,

114

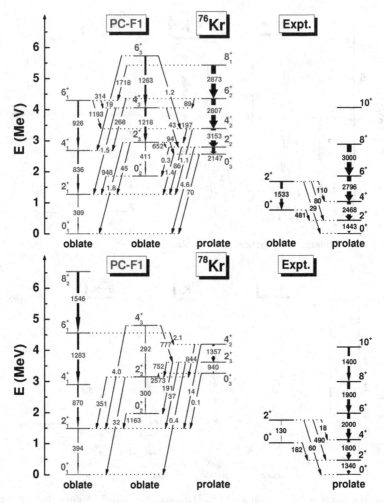

Fig. 3. Same as described in the caption to Fig. 2, but for isotopes ^{76,78}Kr. Experimental data are taken from [7,21,22].

similar as the predication of the non-relativistic calculations in Ref. 13, the ground-state bands of these nuclei are mostly oblate deformed in present calculations, while all are prolate deformed except for ^{72}Kr (not determined yet) according to the measurements.

In conclusion, starting from the relativistic point-coupling energy functional, the spectroscopic properties of low-lying states in ^{72,74,76,78}Kr, including the energy curves, energy spectrum and electric quadrupole transition strengths, have been calculated by mixing of angular momentum projected axially deformed relativistic mean-field states and compared with the corresponding data. The experimental data are qualitatively reproduced, although the ground-state bands of these nuclei are mostly oblate deformed in present calculations, while all (except for ^{72}Kr) are prolate deformed according to the measurements. It has been already pointed out in Ref. 14 that the triaxial degree of freedom might be crucial to reproduce the experimental data. The study including the triaxial deformation with the 3DAMP or the collective Bohr Hamiltonian[23,24] based on the CDF is in progress.

Acknowledgements

This work has been partly supported by the National Natural Science Foundation of China under Grant No. 10947013, the Southwest University Initial Research Foundation Grant to Doctor (SWU109011 and SWU110039) as well as the Fundamental Research Funds for the Central Universities (XDJK2010B007).

References

1. J. L. Wood et al., Phys. Rep. **215**, 101 (1992).
2. R. B. Piercey et al., Phys. Rev. Lett. **47**, 1514 (1981).
3. C. Chandler et al., Phys. Rev. C 56, R2924 (1997).
4. F. Becker et al., Eur. Phys. J. A 4, 103 (1999).
5. E. Bouchez et al., Phys. Rev. Lett. 90, 082502 (2003).
6. E. Poirier, et al., Phys. Rev. C 69, 034307 (2004).
7. E. Clément et al., Phys. Rev. C **75**, 054313 (2007).
8. F. Dickmann et al., Phys. Lett. B **38**, 207 (1972).
9. W. Nazarewicz et al., Nucl. Phys. A **435**, 397 (1985).
10. P. Bonche et al., Nucl. Phys. A **443**, 39 (1985).
11. M. Girod et al., Phys. Rev. Lett. **62**, 2452 (1989).
12. G. A. Lalazissis et al., Nucl. Phys. A **586**, 201 (1995).
13. M. Bender et al., Phys. Rev. C **74**, 024312 (2006).
14. M. Girod et al., Phys. Lett. B **676**, 39 (2009).
15. J. M. Yao et al., Phys. Rev. C **79**, 044312 (2009).
16. J. M. Yao et al., Phys. Rev. C **81**, 044311 (2010).
17. J. M. Yao et al., Phys. Rev. C **83**, 014308 (2011).
18. T. Bürvenich et al., Phys. Rev. C **65**, 044308 (2002).
19. G. de Angelis et al., Phys. Lett. B **415**, 217 (1997).
20. A. Gade et al., Phys. Rev. Lett. **95**, 022502 (2005).
21. A. Görgen et al., Eur. Phys. J. A **26**, 153 (2005).
22. F. Becker et al., Nucl. Phys. A **770**, 107 (2006).
23. T. Nikšić et al., Phys. Rev. C **79**, 034303 (2009).
24. Z. P. Li et al., Phys. Rev. C **79**, 054301 (2009).

ISOSPIN CORRECTIONS FOR SUPERALLOWED β TRANSITIONS

HAOZHAO LIANG[1], NGUYEN VAN GIAI[2], and JIE MENG[3,1,4,*]

[1]*State Key Laboratory of Nuclear Physics and Technology, School of Physics,*
Peking University, Beijing 100871, China
[2]*Institut de Physique Nucléaire, IN2P3-CNRS and Université Paris-Sud,*
F-91406 Orsay Cedex, France
[3]*School of Physics and Nuclear Energy Engineering, Beihang University,*
Beijing 100191, China
[4]*Department of Physics, University of Stellenbosch, Stellenbosch, South Africa*
*E-mail: mengj@pku.edu.cn

The isospin symmetry-breaking corrections δ_c obtained by the self-consistent relativistic RPA calculations are compared with those obtained by the shell model calculations and the isospin- and angular-momentum-projected nuclear density functional theory. It is found that the present theoretical uncertainty of the isospin symmetry-breaking corrections δ_c is of the order of ~ 0.003, rather than ~ 0.0002 as indicated by the shell model calculations. Whether the unitarity of the CKM matrix is fulfilled or not is still an open question.

Keywords: Isospin symmetry-breaking corrections; superallowed β transitions; Cabibbo-Kobayashi-Maskawa matrix.

1. Introduction

The Cabibbo-Kobayashi-Maskawa (CKM) matrix[1,2]

$$\begin{pmatrix} V_{ud} & V_{us} & V_{ub} \\ V_{cd} & V_{cs} & V_{cb} \\ V_{td} & V_{ts} & V_{tb} \end{pmatrix}, \qquad (1)$$

which relates the quark eigenstates of the weak interaction with the quark mass eigenstates, is one of the key transformations in the Standard Model. Kobayashi and Maskawa were awarded the 2008 Nobel Prize in Physics "for the discovery of the origin of the broken symmetry which predicts the existence of *at least* three families of quarks in nature".[3] If the three-generation Standard Model is complete, the rotation embodied in the CKM matrix must be unitary. Therefore, measuring its matrix elements independently and verifying its unitarity condition provide a rigorous test for the Standard Model and set limits on new physics beyond it.

To date, the most precise test is that using the three top-row elements V_{ud}, V_{us} and V_{ub},[4,5] in which the leading term $|V_{ud}|^2$ contributes $\sim 95\%$ to the unitarity sum, thus determining the $|V_{ud}|$ value and restraining its uncertainty is a critical issue for both experimental and theoretical researches.

Four experimental methods to determine the $|V_{ud}|$ value so far include the nuclear $0^+ \to 0^+$ superallowed β transitions,[6] neutron decay,[7] pion β decay[8] and nuclear mirror transitions.[9] In particular, the first method provides the most precise determination.[4,5]

In order to determine the $|V_{ud}|$ value with these nuclear superallowed β transitions, apart from the experimental ft values, the radiative corrections Δ_R^V, δ_R', δ_{NS} and the isospin symmetry-breaking corrections δ_c must also be taken into account.[6] The radiative corrections are due to the emission of real photons and the exchange of virtual photons and Z-bosons in semi-leptonic weak transitions,[10,11] meanwhile, the isospin symmetry-breaking corrections are used to estimate the slight failure of the conserved vector current (CVC) hypothesis caused by the isospin $SU(2)$ symmetry-breaking in finite nuclei.[11] It can be found clearly that, with the radiative and isospin symmetry-breaking corrections, the corrected $\mathcal{F}t$ should be nucleus-independent.

The isospin symmetry-breaking corrections δ_c characterize the slight reduction of the superallowed transition strengths $|M_F|^2$ from the ideal value $|M_0|^2$:

$$|M_F|^2 = |\langle f| T_\pm |i\rangle|^2 = |M_0|^2(1 - \delta_c), \tag{2}$$

with $M_0 = \sqrt{2}$ for $T = 1$ states having exact isospin symmetry. Shell model calculations[11] are generally used to evaluate the corrections δ_c for the past several decades. However, recently it was pointed out that the significant radial excitations are neglected in the treatment used in Ref. 11, thus the corrections δ_c therein are overestimated,[12] which leads to the overestimate of the $|V_{ud}|$ value. Alternatively, the self-consistent Random Phase Approximation (RPA) based on microscopic mean field theories is another reliable approach for evaluating the superallowed transition strength $|M_F|^2$. Such calculations have been performed for several nuclei with the non-relativistic Skyrme Hartree-Fock approach in Ref. 13. During the last decade, great efforts have been dedicated to developing the charge-exchange (Q)RPA within the relativistic framework.[14–16] In particular, a fully self-consistent charge-exchange RPA based on the relativistic Hartree-Fock (RHF) approach[17] was established.[16] A very satisfactory description of spin-isospin resonances was obtained without any readjustment of the energy functional. Adopting these self-consistent relativistic RPA approaches, the corrections δ_c have been systematically investigated in Refs. 18 and 19. Recently, the evaluations of the corrections δ_c have also been performed within the isospin- and angular-momentum-projected nuclear density functional theory (DFT). The preliminary results are presented in Ref. 20.

In this report, the isospin symmetry-breaking corrections δ_c obtained by the self-consistent relativistic RPA calculations will be compared with those obtained by the shell model calculations as well as by the projected DFT. Together with the most updated $|V_{us}|$ and $|V_{ub}|$ values, the unitarity of the CKM matrix will also be discussed.

2. Discussion

Table 1. Isospin symmetry-breaking corrections δ_c expressed in %. Details are given in the text.

	PKO1[18]	T&H[11]	SV[20]
^{10}C $\to$ ^{10}B	0.082	0.175(18)	0.559(56)
^{14}O $\to$ ^{14}N	0.114	0.330(25)	0.303(30)
^{18}Ne $\to$ ^{18}F	0.270	0.565(39)	
^{22}Mg $\to$ ^{22}Na		0.380(22)	0.243(24)
^{26}Si $\to$ ^{26}Al	0.176	0.435(27)	
^{30}S $\to$ ^{30}P	0.497	0.855(28)	
^{34}Ar $\to$ ^{34}Cl	0.268	0.665(56)	0.865(87)
^{38}Ca $\to$ ^{38}K	0.313	0.765(71)	
^{42}Ti $\to$ ^{42}Sc	0.384	0.935(78)	
^{26}Al^m $\to$ ^{26}Mg	0.139	0.310(18)	0.494(49)
^{34}Cl $\to$ ^{34}S	0.234	0.650(46)	0.679(68)
^{38}K^m $\to$ ^{38}Ar	0.278	0.655(59)	
^{42}Sc $\to$ ^{42}Ca	0.333	0.665(56)	0.767(77)
^{46}V $\to$ ^{46}Ti		0.620(63)	0.759(76)
^{50}Mn $\to$ ^{50}Cr		0.655(54)	0.740(74)
^{54}Co $\to$ ^{54}Fe	0.319	0.770(67)	0.671(67)
^{62}Ga $\to$ ^{62}Zn		1.48(21)	0.925(93)
^{66}As $\to$ ^{66}Ge	0.475	1.56(40)	
^{70}Br $\to$ ^{70}Se	1.140	1.60(25)	
^{74}Rb $\to$ ^{74}Kr	1.088	1.63(31)	2.06(21)

In Table 1, the isospin symmetry-breaking corrections δ_c for the $0^+ \to$ 0^+ superallowed transitions obtained by self-consistent RHF+RPA calcu-

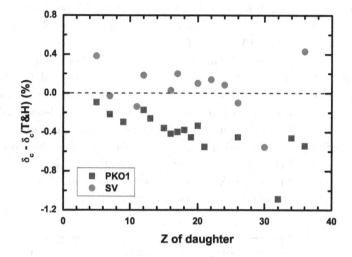

Fig. 1. (Color online) Differences of the isospin symmetry-breaking corrections δ_c as a function of the charge Z of the daughter nucleus. The T&H results are taken as references, while the uncertainties of δ_c are not taken into account.

lations[18] with PKO1 parametrization[17] are compared with those obtained by the shell model calculations (T&H)[11] and the isospin- and angular-momentum-projected nuclear density functional theory[20] with SV Skyrme parametrization.[21] In order to clearly see the different predictions among these three approaches, the differences of the isospin symmetry-breaking corrections δ_c are shown in Fig. 1 as a function of the charge Z for the daughter nucleus, where the T&H results are taken as references, while the uncertainties of δ_c are not taken into account.

On one hand, it is found that the RHF+RPA results are systematically smaller than those of T&H by around 0.4% for most cases. Since it was pointed out that the corrections δ_c obtained by the shell model calculations may be overestimated due to the neglect of the radial excitations, it is worthwhile to notice this systematical discrepancy for the future investigations. Meanwhile, it should be also pointed out that, in the present relativistic RPA calculations, the effects of pairing correlations and deformation are not taken into account. In addition, the missing charge symmetry breaking (CSB) and charge independence breaking (CIB) forces are expected to enlarge the corrections δ_c by $\sim 0.1\%$ according to the calculations in Ref. 13.

On the other hand, it is found that the projected DFT results are similar to the T&H results in average, while the individual values differ from each other by the amount of $0.1 \sim 0.2\%$ for most cases.

Three pf-shell nuclei – ^{62}Ga, ^{66}As, and ^{74}Rb – should be paid more attention to. In particular, the uncertainty of the experimental ft value for ^{62}Ga has become relatively small.[6] Since the weight $\omega_i = 1/\sigma_i^2$ is used to calculate the chi-square per degree of freedom χ^2/ν,[4] the underestimate (or overestimate) of the correction δ_c for such nucleus leads to one of the dominant contributions to the value of χ^2/ν, which is crucial to determine the confidence level of the nuclear-structure-dependent corrections. Therefore, the nuclear-structure-dependent corrections for ^{62}Ga could become one of the critical cases in the future theoretical investigations.

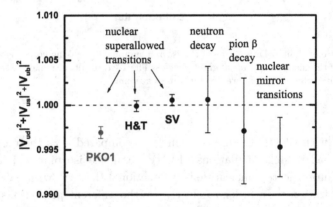

Fig. 2. (Color online) The sum of squared top-row elements of the CKM matrix.

Combining the $|V_{ud}| = 0.97273(27)$ (PKO1),[18] $|V_{ud}| = 0.97425(22)$ (H&T)[6] or $|V_{ud}| = 0.97459(24)$ (SV),[20] and the most updated values of other two top-row matrix elements $|V_{us}| = 0.2252(9)$ and $|V_{ub}| = 0.00389(44)$,[4] the sum of squared top-row elements of the CKM matrix extracted from the superallowed β transitions is shown in Fig. 2. These results are also compared with the most updated values obtained in neutron decay,[4] pion β decay[4] and nuclear mirror transitions.[9] It is interesting to note that whether the unitarity of the CKM matrix is fulfilled or not is still an open question. Remarkably, adopting different theoretical framework, the theoretical uncertainty of the isospin symmetry-breaking corrections δ_c is of the order of magnitude ~ 0.003, rather than ~ 0.0002 as it was concluded from the shell model calculations in Ref. 5.

3. Summary

The isospin symmetry-breaking corrections δ_c obtained by the self-consistent relativistic RPA calculations are compared with those obtained by the shell model calculations as well as by the isospin- and angular-momentum-projected nuclear density functional theory. It is pointed out that the present theoretical uncertainty of the isospin symmetry-breaking corrections δ_c is of the order of magnitude ~ 0.003, rather than ~ 0.0002 as suggested by the shell model calculations. Thus, in our opinion, whether the unitarity of the CKM matrix is fulfilled or not is still an open question.

Acknowledgments

This work was partly supported by the Major State 973 Program 2007CB815000, the NSFC under Grant No. 10975008, and China Post-doctoral Science Foundation Grant No. 20100480149.

References

1. N. Cabibbo, *Phys. Rev. Lett.* **10**, 531 (1963).
2. M. Kobayashi and T. Maskawa, *Prog. Theor. Phys.* **49**, 652 (1973).
3. The Nobel Prize in Physics 2008, http://nobelprize.org/nobel_prizes/physics/laureates/2008/.
4. K. Nakamura *et al.*, *J. Phys. G* **37**, 075021 (2010).
5. I. S. Towner and J. C. Hardy, *Rep. Prog. Phys.* **73**, 046301 (2010).
6. J. C. Hardy and I. S. Towner, *Phys. Rev. C* **79**, 055502 (2009).
7. D. Thompson, *J. Phys. G* **16**, 1423 (1990).
8. D. Počanić *et al.*, *Phys. Rev. Lett.* **93**, 181803 (2004).
9. O. Naviliat-Cuncic and N. Severijns, *Phys. Rev. Lett.* **102**, 142302 (2009).
10. W. J. Marciano and A. Sirlin, *Phys. Rev. Lett.* **96**, 032002 (2006).
11. I. S. Towner and J. C. Hardy, *Phys. Rev. C* **77**, 025501 (2008).
12. G. A. Miller and A. Schwenk, *Phys. Rev. C* **80**, 064319 (2009).
13. H. Sagawa, N. Van Giai and T. Suzuki, *Phys. Rev. C* **53**, 2163 (1996).
14. C. De Conti, A. P. Galeão and F. Krmpotić, *Phys. Lett. B* **444**, 14 (1998).
15. N. Paar, T. Nikšić, D. Vretenar and P. Ring, *Phys. Rev. C* **69**, 054303 (2004).
16. H. Liang, N. Van Giai and J. Meng, *Phys. Rev. Lett.* **101**, 122502 (2008).
17. W. H. Long, N. Van Giai and J. Meng, *Phys. Lett. B* **640**, 150 (2006).
18. H. Liang, N. Van Giai and J. Meng, *Phys. Rev. C* **79**, 064316 (2009).
19. H. Liang, N. Van Giai and J. Meng, *J. Phys. Conf. Ser.* **205**, 012028 (2010).
20. W. Satuła, J. Dobaczewski, W. Nazarewicz and M. Rafalski (2011), arXiv:1101.0939v1 [nucl-th].
21. M. Beiner, H. Flocard, N. Van Giai and P. Quentin, *Nucl. Phys. A* **238**, 29 (1975).

THE POSITIVE-PARITY BAND STRUCTURES IN ^{108}Ag

C. LIU, S. Y. WANG*, B. QI, D. P. SUN, C. J. XU, L. LIU, B. WANG, X. C. SHEN,
M. R. QIN, H. CHEN

Shandong Provincial Key Laboratory of Optical Astronomy and Solar-Terrestrial Environment, School of Space Science and Physics, Shandong University at Weihai, Weihai 264209, People's Republic of China
E-mail: sywang@sdu.edu.cn

L. H. ZHU, X. G. WU, G. S. LI, C. Y. HE, Y. ZHENG, L. L. WANG, B. ZHANG

China Institute of Atomic Energy, Beijing 102413, People's Republic of China

G. Y. LIU, Y. W. WANG

Department of Physics, Jilin University, Changchun 130023, People's Republic of China

The high-spin states in the odd-odd nucleus ^{108}Ag were populated in the reaction ^{104}Ru(^{7}Li,3n) at a beam energy of 33 MeV. The previously known positive-parity band structures have been extended to higher spins, and their configurations have been discussed. Alignments, band crossing frequencies, and $B(M1)/B(E2)$ ratios have been analyzed in the framework of the cranking model.

Keywords: High spin states; rotational bands; configuration assignment.

1. Introduction

The low-lying states of ^{108}Ag had been studied through atomic beam resonance,[1] and the ground state has been assigned $I^{\pi} = 1^{+}$. Two isomeric states $I^{\pi} = 3^{+}$ and $I^{\pi} = 6^{+}$ were also identified from the β radioactive decay of Pd and Cd isotopes.[2] Recently, high-spin states of ^{108}Ag had been studied by Espinoza-Quiñones *et al.* through in-beam spectroscopy.[3] However, Ref.[3] pointed out that there are some puzzling problems about the configuration assignment of the positive-parity bands. This motivated our study of the positive-parity band structures in ^{108}Ag.

2. Experimental procedure

High spin states in ^{108}Ag were populated by using the ^{104}Ru(^{7}Li, 3n) reaction. The experiment was carried out at China Institute of Atomic Energy (CIAE). The tandem accelerator provided a 33 MeV ^{7}Li beam which bombarded a 2 mg/cm^2 Ru target backed with 10 mg/cm^2 Pb. γ-rays from the reaction were detected using an array of 12 BGO suppressed HPGE detectors and a Clover detector. A total of 2×10^8 γ-γ coincidence events were collected in event-by-event mode. The coincidence data were recalibrated to 0.5 keV/channel and sorted into a 4096×4096 symmetrized matrix as well as into an asymmetric directional correlation ratios of oriented states (DCO) matrix in the off-line analysis.

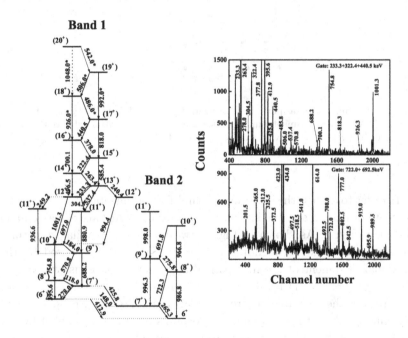

Fig. 1. Partial level scheme of ^{108}Ag (left) deduced from the present work and the sample coincidence spectra (right) supporting this level scheme.

3. Results and discussion

The level scheme for ^{108}Ag deduced from the present work and the typical examples of coincidence spectra are presented in Fig. 1. The placements of γ-rays in the positive parity band structures are in agreement with those of

Ref.[3] except that six new transitions have been added. All γ-rays reported in this work feed eventually toward an isomeric state at $I^\pi = 6^+$. This 6^+ isomeric state has been interpreted as $[\pi(g_{9/2})^{-3} \otimes \nu d_{5/2}]_{6^+}$ coupled to the ^{110}Sn core.[4] Band 2 was built on this isomer, and assigned to the $\pi g_{9/2} \otimes \nu d_{5/2}$ configuration.

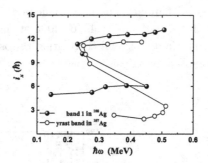

Fig. 2. Experimental quasiparticle alignments i_x for band 1 in ^{108}Ag and the yrast band in ^{107}Ag. The experimental quasiparticle alignments were extracted with the Harris parameters of $J_0 = 3.6\hbar^2$ MeV^{-1} and $J_1 = 29.8\hbar^4$MeV^{-3}.

In order to discuss the structure of band 1, experimental aligned angular momenta of band 1 and the known $\pi g_{9/2}$ band of ^{107}Ag[5] are extracted and plotted in Fig. 2. As shown in Fig. 2, two bands have similar structures and show the same backbending frequency ($\sim$0.35 MeV) caused by the rotational alignment of the first pair of $h_{11/2}$ neutrons. Thus, we suggest that a $\pi g_{9/2}$ orbital is involved in the configurations of band 1. In this mass region, the available neutron orbitals are $h_{11/2}$, $d_{5/2}$, $g_{7/2}$, $s_{3/2}$ and $s_{1/2}$, among which the $h_{11/2}$ and $d_{5/2}$ orbitals have been occupied by the yrast band and band 2 in ^{108}Ag, respectively. In addition, the $d_{3/2}$ and $s_{1/2}$ orbitals are far from the Fermi surface of ^{108}Ag. Therefore, the $g_{7/2}$ is the most possible neutron orbital for band 1. Taking the above information into account, we propose a $\pi g_{9/2} \otimes \nu g_{7/2}$ configuration before the backbending, a $\pi g_{9/2} \otimes \nu g_{7/2}(\nu h_{11/2})^2$ configuration after the backbending for band 1.

The experimental $B(M1)/B(E2)$ ratios are extracted and compared with those calculated for different possible configurations of bands 1 and 2 using the geometric model of Dönau,[6] as given in Fig. 3. This comparison favors the present configuration assignment of bands 1 and 2.

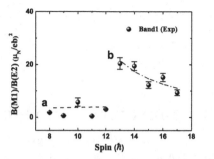
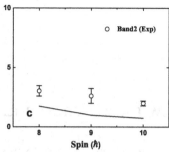

Fig. 3. Experimental $B(M1)/B(E2)$ ratios for band 1 and band 2 in ^{108}Ag. The solid curves corresponds to calculated $B(M1)/B(E2)$ ratios. a) $\pi g_{9/2} \otimes \nu g_{7/2}$, b) $\pi g_{9/2} \otimes \nu g_{7/2}(\nu h_{11/2})^2$, c) $\pi g_{9/2} \otimes \nu d_{5/2}$. Parameters used in the calculations: $Q_0 \approx 2$ eb, $g_R = Z/A \approx 0.44$, $g_p(g_{9/2}) = 1.260$, $g_n(g_{7/2}) = 0.255$, $g_n(d_{5/2}) = -0.459$, $i_p(g_{9/2}) = 2.37$,[5] $i_n(g_{7/2}) = 2.25$,[7] $i_n(d_{5/2}) = 2.25$,[7] $i_{nn}(h_{11/2}) = 9$.[5] The proton and neutron g factors are determined from $g_{p(n)} = g_l + (g_s - g_l)/(2l + 1)$ with $g_l = 1(0)$ for proton (neutron) and $g_s = 0.6 g_s$(free).

4. Conclusion

High spin states in ^{108}Ag have been studied via the ^{104}Ru(^{7}Li, 3n) reaction. One previously known positive-parity band has been extended up to 20^+. Alignments, band crossing frequencies, and $B(M1)/B(E2)$ ratios have been analyzed in the framework of the cranking model. we suggested a $\pi g_{9/2} \otimes \nu g_{7/2}$ configuration before the backbending, a $\pi g_{9/2} \otimes \nu g_{7/2}(\nu h_{11/2})^2$ configuration after the backbending for band 1 and a $\pi(g_{9/2}) \otimes \nu d_{5/2}$ configuration for band 2.

Acknowledgements

This work is supported by the National Natural Science Foundation (Grant Nos. 10875074 and 11005069), the Shandong Natural Science Foundation (Grant No. ZR2010AQ005), and the Major State Research Development Programme (No. 2007CB815005) of China.

References

1. G. K. Rochester, and K. F. Smith, *Phys. Lett.* **8**, 266 (1964).
2. O. C. Kistener, and A. W. Sunyar, *Phys. Rev.* **143**, 918 (1966).
3. F. R. Espinoza-Quiñones *et al.*, *Phys. Rev. C* **52**, 104 (1995).
4. T. Hattori *et al.*, *J. Phys. Soc. Jpn.* **41**, 1830 (1976).
5. F. R. Espinoza-Quiñones *et al.*, *Phys. Rev. C* **55**, 3 (1997).
6. F.Dönau, *Nucl. Phys. A* **471**, 469 (1987).
7. K. R. Pohl *et al.*, *Phys. Rev. C* **55**, 2682 (1996).

NEW BAND STRUCTURES IN ODD-ODD ^{120}I AND ^{122}I

LIU GONGYE, LI LI, LI XIANFENG, MA YINGJUN*, ZHAO YANXIN
ZHOU WENPING, YANG DONG, MA KEYAN, LI CONGBO, YANG YANJI,
YU DEYANG, LU JINGBIN

Physics Department, Jilin University, QianJin Street 2699
Changchun, 130012, China
*E-mail: myj@jlu.edu.cn

WU XIAOGUANG, ZHU LIHUA, HE CHUANGYE, ZHENG YUN, WANG LIELIN,
HAO XIN, LI GUANGSHENG

China Institute of Atomic Energy, Box 275-10
Beijing, 102413, China

High spin states of the odd-odd ^{120,122}I isotopes have been investigated via the ^{110}Pd (^{14}N, 4n), ^{114}Cd (^{10}B, 4n) and ^{116}Cd (^{11}B, 5n) reactions at beam energies of 64, 48 and 68 MeV. In ^{120}I, the previously known bands are extended, and a new band structure has been established. In ^{122}I, the yrast band is extended up to (29$^+$)and assigned to the $[\pi h_{11/2}(\pi g_{7/2})^2]_{23/2-} \otimes [(\nu h_{11/2})^3(\nu d_{5/2})^2]_{35/2-}$ configuration, which corresponds to the full alignment of all the valance nucleons outside the semi-closed shell.

Keywords: High spin states; band structure; doubly odd nucleus.

1. Introduction

This work is a part of a program for systematic study of iodine nuclei with $A \sim 120$. In our earlier work, the iodine isotopes ^{123}I[1] and ^{126}I[2] had been investigated. In the current work, we focus our attention to the odd-odd nuclei ^{120,122}I. In the previous work by H. Kaur *et al.*,[3,4] these nuclei were studied via the ^{108}Pd (^{16}O, 4n),^{114}Cd (^{11}B, 5n) and ^{116}Cd (^{11}B, 5n) reactions at 84, 60 and 64 MeV. Three band structures were respectively reported in ^{120}I and ^{122}I, and the yrast ones in both nuclei were assigned to the $\pi(g_{7/2}/d_{5/2}) \otimes \nu h_{11/2}$ configuration. Furthermore, a loss in collectivity in the yrast bands of the two nuclei was observed around spin $I = 21$. Later, several new band structures in ^{122}I were established by C.-B. Moon *et al.*,[5] through the ^{7}Li induced reaction, and a level scheme of ^{122}I superseding the result by H. Kaur *et al.*, has been compiled into Nuclear Data Sheets.[6] It is worthwhile to emphasize that there is a serious discrepancy between Refs. 3 and 5,6 about the coincidence structure and configuration assignment of the yrast band in ^{122}I.

2. Experiment details

The excited states of ^{120}I and ^{122}I nuclei have been studied by using in-beam gamma-ray spectroscopy with the ^{110}Pd (^{14}N, 4n), ^{114}Cd (^{10}B, 4n) and ^{116}Cd (^{11}B, 5n) reactions at beam energies of 64, 48 and 68 MeV. The beams were provided by the HI-13 tandem accelerator of China Institute of Atomic Energy (CIAE) in Beijing. The ^{110}Pd was a metallic foil of 2.4 mg/cm^2 thickness with 0.2 mg/cm^2 Au backing. Targets of ^{114}Cd and ^{116}Cd were self-supporting foils with the thickness of 5 mg/cm^2 and 4.8 mg/cm^2 . A detection array consisting of 12 Compton-suppressed HPGe detectors and 2 planar HPGe detectors was used for $\gamma - \gamma$ coincidence measurement in the experiments of ^{120}I. For ^{122}I, a detection array consisting of 10 Compton-suppressed HPGe detectors, 2 planar HPGe detectors and a CLOVER detector was used. About 200, 170 and 500 million coincidence events were accumulated for the ^{110}Pd (^{14}N, 4n), ^{114}Cd (^{10}B, 4n) and ^{116}Cd (^{11}B, 5n) reactions, respectively.

In the off-line data analysis, the coincidence data were recalibrated to 0.5 keV/channel and sorted into several symmetric 4096 × 4096 matrices. Non-symmetric (ADO sorting) matrices were also created for the purpose of extracting multipolarity information of the γ-rays using the method of ADO ratios. Background-subtracted coincidence spectra were generated and intensity analysis was performed using the PC-base program MXA.

3. Results and discussions

3.1. *New structures in* ^{120}I

A partial level scheme deduced from the present work is shown in Fig. 1, where the transitions have been arranged into bands labeled 1 to 4 for the convenience of discussions. The placement of γ-transitions into the level scheme is based on their coincidence relationships, energy sums and relative intensities. In comparison with the previous study by H. Kaur *et al.*,[4] we have identified two new bands, namely, bands 1 and 4. Furthermore, the 992 keV transition placed on the top of band 2 in Ref. 4 has now been placed as a linking transition from band 1 to band 2, and a new transition with energy of 905 keV has been placed on the top of band 2. In addition, the ordering of the 158.8 and 199.4 keV transitions below bands 1 and 2 has been exchanged. Bands 3 and 4 were also reported in a recent study by C.-B. Moon *et al.*[7] Our result is in agreement with theirs and band 4 has been extended up by two units of spin.

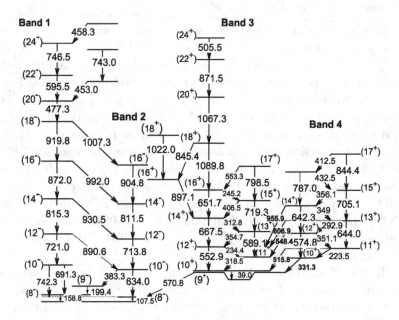

Fig. 1.   Partial level scheme of ^{120}I deduced from the present work.

3.2. *Yrast band in* ^{122}I

Figure 2 shows the structure of the yrast band of ^{122}I. Three new transitions with energies of 1101, 1161 and 248 keV have been added onto the top of the yrast band. For the lower part of the yrast band, the coincidence relationships among the different γ-transitions deduced from this work agree with those reported in Refs. 5,6 and disagree with those in Ref. 3 . C.-B. Moon *et al.* assigned the $\pi h_{11/2} \otimes \nu h_{11/2}$ configuration to the yrast band whereas H. Kaur *et al.* assigned the $\pi(g_{7/2}/d_{5/2}) \otimes \nu h_{11/2}$ configuration. Based on considerations from systematics, we adopt the assignment of C.-B. Moon *et al.* Spin and parity assignments for the lower levels in the yrast band are also taken from Refs. 5,6, and for higher levels, they are made on the basis of the present measurement of ADO ratios. The spin of the $I^\pi = (29^+)$ state can be generated from the $[\pi h_{11/2}(\pi g_{7/2})^2]_{23/2^-} \otimes [(\nu h_{11/2})^3(\nu d_{5/2})^2]_{35/2^-}$ configuration. Such a configuration corresponds to a state where all valance nucleons outside the closed or semi-closed shell align their single-particle angular momenta along the symmetric axis of the nucleus.

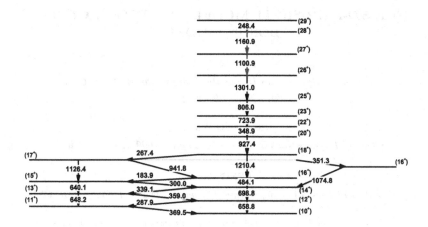

Fig. 2. Partial level scheme showing the yrast band in ^{122}I as deduced from the present work.

Acknowledgments

This work was supported by National Natural Science Foundation of China under Grant No.10675053 and No.10475033, and National Basic Research Programm of China under Grant No.2007CB815005.

References

1. Y. X. Zhao, T. Komatsubara and Y. J. Ma, *Chin. Phys. Lett.* **26**, 082301 (2009).
2. R. J. Li, *High Energy Phys. And Nucl. Phys.* **29**, 1 (2005).
3. H. Kaur and J. Singh, *Phys. Rev. C* **45**, 2234 (1997).
4. H. Kaur and J. Singh, *Phys. Rev. C* **55**, 512 (1997).
5. C. B. Moon, *J. Korean Phys. Soc.* **43**, S100 (2003).
6. T. Tamura, *Nucl. Dat. Sheets* **108**, 455 (2007).
7. C. B. Moon, *J. Korean Phys. Soc.* **44**, 244 (2003).

THE SD-PAIR SHELL MODEL AND INTERACTING BOSON MODEL

YAN-AN LUO

School of Physics, Nankai University, Tianjin 300071, P. R. China
email:luoya@nankai.edu.cn

FU-YONG WANG

School of Physics, Nankai University, Tianjin 300071, P. R. China

FENG PAN

Department of Physics, Liaoning Normal University, Dalian 116029, P. R. China

Department of Physics and Astronomy, Louisiana State University,
Baton Rouge, LA 70803, USA

JERRY P. DRAAYER

Department of Physics and Astronomy, Louisiana State University,
Baton Rouge, LA 70803, USA

PING-ZHI NING

School of Physics, Nankai University, Tianjin 300071, P. R. China

The SD-pair shell model (SDPSM) is shown to reproduce approximately typical spectra, $E2$ transition strengths of the $U(5)$, $SO(6)$, $SU(3)$ and $SU^*(3)$ limits of the interacting boson model (IBM). The shape phase transitional patterns of the IBM can also be reproduced in the SDPSM. This analysis confirms that the IBM has a sound shell-model foundation; it also demonstrates that the truncation scheme adopted in the SDPSM is reasonable.

Keywords: SD-pair shell model; interacting boson model; shape phase transition.

1. Introduction

By using the generalized Wick theorem for fermion clusters,[1] the nucleon-pair shell model (NPSM) has been proposed for nuclear collective motion in which collective nucleon pairs with various angular momenta serve as the building blocks.[2,3] The NPSM has the advantages that it is flexible enough to include the broken pair approximation,[4] the pseudo SU(2) or the favored pair model,[5] and the fermion dynamical symmetry model (FDSM)[6] as special cases, and it allows various truncation schemes ranging from the truncation to the S-D subspace up to the full shell-model space.

Because the computational time increases dramatically with the size of the subspace, one normally truncates this shell-model space for medium and heavy mass nuclei to the collective S-D subspace. The latter is called the SD-pair shell model (SDPSM). In the SDPSM the Hamiltonian is diagonalized exactly in the S-D space. Since the interacting boson model(IBM) was also in the S-D space, it is interesting to study whether the SDPSM can reproduce the similar results as those of the IBM, and this is the aim of this paper.

The paper is organized as follows. Section II is devoted to the discussion of the limiting cases of the IBM, the nuclear shape phase transitional pattern of the IBM will be presented in section III. Section IV is a short summary of the results.

2. The dynamical limits of the IBM

The vibrational, rotational, γ-unstable and triaxially deformed spectra corresponding to those of $U(5)$, $SU(3)$, $SO(6)$ and $SU^*(3)$ limits in the IBM can all be reproduced in the SDPSM. As an example, the results corresponding to the triaxially deformed limit $SU^*(3)$ in the IBM were given here. The detailed discussion about the limiting cases can be found in Refs.[7,8]

To see if the SDPSM can produce the properties corresponding to the $SU^*(3)$, a schematic Hamiltonian like that from Ref.[9] is used,

$$H(SU^*(3)) = -\kappa_\pi Q^{(2)}(\pi) \cdot Q^{(2)}(\pi) - \kappa_\nu Q^{(2)}(\nu) \cdot Q^{(2)}(\nu)$$
$$+ \kappa Q^{(2)}(\pi) \cdot Q^{(2)}(\nu) \tag{1}$$

where

$$Q^{(2)} = \sum_i r_i^2 Y^2(\theta_i, \phi_i) \tag{2}$$

and κ_π, κ_ν and κ are the quadrupole-quadrupole interaction strength between protons and protons, neutron and neutron, and protons and neutron, respectively.

The $E2$ transition operator is given within this framework as

$$T(E2) = e_\pi Q_\pi^{(2)} + e_\nu Q_\nu^{(2)}, \tag{3}$$

where e_π and e_ν are effective charges of the proton and neutron, respectively.

The collective S-pair is defined as

$$S^\dagger = \sum_a y(aa0)(C_a^\dagger \times C_a^\dagger)^0 \qquad (4)$$

Because we use the degenerate single-particle levels, the S-pair structure coefficient is fixed to be $y(aa0) = \hat{a}\frac{v_a}{u_a} = \hat{a}\sqrt{\frac{N}{\Omega_a - N}}$, where Ω_a is defined as $\Omega_a = a + 1/2$ and N is the number of pairs for like-nucleons. The D-pair is obtained by using the commutator

$$D^\dagger = \tfrac{1}{2}[Q^{(2)}, S^\dagger] = \sum_{ab} y(ab2)\left(C_a^\dagger \times C_b^\dagger\right)^2. \qquad (5)$$

After symmetrization, it is easy to show that

$$y(ab2) = -\frac{1}{2}q(ab2)\left[\frac{y(aa0)}{\hat{a}} + \frac{y(bb0)}{\hat{b}}\right]. \qquad (6)$$

In Ref.[9] it was shown that the triaxial rotor can be realized for the case with protons as particles (or holes) and neutrons as holes (or particles). That is, the protons and neutrons have opposite intrinsic quadrupole deformation. In this paper we assume that protons are particle-like, while neutrons are hole-like. Therefore, the proton-neutron coupled system with $N_\pi = \bar{N}_\nu = 5$ was studied. We restricted ourselves to the gds shell with the same set of orbits taken for both protons and neutron-holes. The quadrupole-quadrupole interaction strengths were fixed at $\kappa_\pi = \kappa_\nu = \kappa = 0.01$ MeV$/r_0^4$ for simplicity.

The triaxial pattern we obtained can be appreciated by comparing the low-lying energy levels of the SDPSM shown in Fig. 1 with those of the triaxial deformed spectra of the IBM.[9–11] For example, the two 2^+ states, three 4^+ states, etc., are almost degenerate, and the energy ratios $R_{32} = E_{3_1^+}/(E_{2_1^+} + E_{2_2^+})$ and $R_{35} = (E_{5_1^+} + E_{5_2^+})/E_{3_1^+}$ are 1.01 and 4.91, respectively, which fall very close to the $SU^*(3)$ values of 1.0 and 5.0.[9]

To track the rotational pattern, we also show the energies of the ground and quasi-γ bands plotted as a function of $J(J+1)$ in Fig. 2. The linearity of the results show that these bands vary as $J(J + 1)$, which is a clear signature of their rotational character.

It is well known that $B(E2)$ values can also be used to test the collectivity of the low-lying states. To further explore how well the nature of the $SU^*(3)$ limit of the IBM can be produced within the SDPSM, the relative $B(E2)$ values were calculated with the results given in Table 1. The empirical relations that $e_\pi = 1 + e_\nu$ and $e_\nu = Z/A$ have been used for obtaining the numerical values for effective charges. In this letter, $A = 100$ and $Z = 50$ were assumed, and we got $e_\pi = 1.5e$ and $e_\nu = 0.5e$. Because we choose the hole-like picture for the neutron SD subspace, the operator $Q_\nu^{(2)}$ changes its sign due to the particle-hole transformation, a negative effective charge for neutron-hole was used.[12] Therefore, the effective charges we used are $e_\pi = -3e_\nu = 1.5e$. In order to make a comparison with those of the IBM, the results taken from[10] are also given. It is seen that the SDPSM can reproduce the relative $B(E2)$ values of the $SU^*(3)$ limit in the IBM very well.

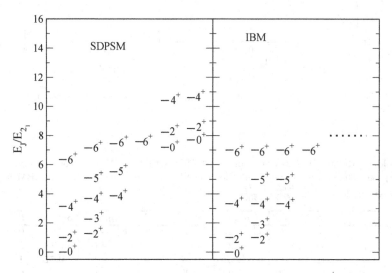

Fig. 1. Some low-lying energy ratios between exciting states and 2_1^+ states in the SDPSM and IBM.

3. The shape phase transitional pattern in the SDPSM

Recently, the nuclear shape phase transition have been studied extensively. To see whether the SDPSM can reproduce the similar results as those of the IBM, the shape phase transitional pattern in the IBM are also studied in the SDPSM. It is found that the SDPSM can reproduce the similar results as those of the IBM. As an example, the shape phase transitional pattern between $U(5)$ and $SU(3)$ was given here. The detailed discussion can be found in Refs.[13,14]

134

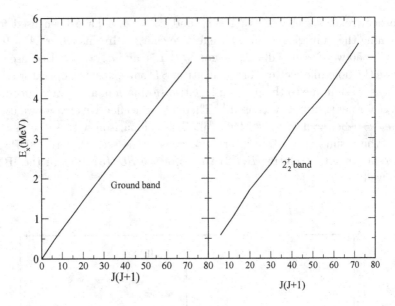

Fig. 2.   The ground and quasi-γ bands with $J(J+1)$ in the SDPSM.

Table 1.   Relative $B(E2)$ values for the SDPSM with $\bar{N}_\nu = N_\pi = 5$. For comparative purposes, the corresponding results for the $SU^*(3)$ symmetry limit of the IBM are also given. $R_{B(E2)} \equiv \frac{B(E2;J_i \rightarrow J_f)}{B(E2;2_1 \rightarrow 0_1)}$.

$R_{B(E2)}$	SDPSM	IBM	$R_{B(E2)}$	SDPSM	IBM	$R_{B(E2)}$	SDPSM	IBM
$4_1^+ \rightarrow 2_1^+$	1.36	1.35	$4_3^+ \rightarrow 4_2^+$	1.63	1.77	$6_2^+ \rightarrow 4_2^+$	0.68	0.63
$6_1^+ \rightarrow 4_1^+$	1.58	1.61	$4_3^+ \rightarrow 3_1^+$	0.59	0.53	$6_2^+ \rightarrow 6_1^+$	0.16	0.13
$2_2^+ \rightarrow 2_1^+$	1.42	1.44	$5_1^+ \rightarrow 3_1^+$	0.87	0.89	$6_2^+ \rightarrow 5_2^+$	0.71	1.02
$3_1^+ \rightarrow 2_2^+$	1.77	1.79	$5_1^+ \rightarrow 4_2^+$	0.92	0.95	$6_3^+ \rightarrow 6_2^+$	0.83	0.53
$4_2^+ \rightarrow 2_2^+$	0.56	0.56	$5_2^+ \rightarrow 5_1^+$	0.45	0.46	$6_3^+ \rightarrow 5_1^+$	0.10	0.08
$4_1^+ \rightarrow 3_1^+$	0.70	0.79	$5_2^+ \rightarrow 4_3^+$	1.51	1.55	$6_3^+ \rightarrow 4_3^+$	0.39	0.39
$4_2^+ \rightarrow 4_1^+$	0.34	0.26				$6_4^+ \rightarrow 5_2^+$	0.90	0.65

In the shell model description, the pairing and quadrupole-quadrupole interactions are the most important short-range and long-range correlations. In considering that the Hamiltonian used to study the shape phase transition in the IBM is mainly composed of the monopole pairing and quadrupole-quadrupole interaction (e.g., Ref.[15,16]), a schematic Hamiltonian is adopted in the SDPSM, which is a combination of the monopole pairing and quadrupole-quadrupole interaction with

$$H_X = \sum_{\sigma=\pi,\nu} (-G_\sigma S_\sigma^\dagger S_\sigma - \kappa_\sigma Q_\sigma^{(2)} \cdot Q_\sigma^{(2)}) - \kappa Q_\pi^{(2)} \cdot Q_\nu^{(2)}, \qquad (7)$$

$$S^\dagger = \sum_a \frac{\widehat{a}}{2} (C_a^\dagger \times C_a^\dagger),$$

$$Q^{(2)} = \sqrt{16\pi/5} \sum_i r_i^2 Y^2(\theta_i, \phi_i)$$

where X in H_X is denoted as $U(5)$, $SU(3)$, or $SO(6)$ corresponding to vibrational, rotational, or gamma-soft limiting case in the model, G_σ and κ_σ are the pairing and quadrupole-quadrupole interaction strength between identical-nucleons, respectively. κ is the quadrupole-quadrupole interaction strength between proton and neutrons. In this paper, we set $G_\pi = G_\nu$ and $\kappa_\pi = \kappa_\nu$.

To study the phase transitional patterns, the Hamiltonian for proton-neutron coupled system is written as

$$H = (1 - \alpha)H_{U(5)} + \alpha H_{SU(3)}, \qquad (8)$$

where $0 \leq \alpha \leq 1$ is a control parameter.

The $E2$ transition operator adopted is

$$T(E2) = e_\pi Q_\pi^{(2)} + e_\nu Q_\nu^{(2)}, \qquad (9)$$

where $e_\pi(e_\nu)$ is the effective charge for proton (neutron).

To study the vibration-rotation phase transition, a system with $N_\pi = N_\nu = 3$ in gds shell was used. The methods used to determine the structure of S-D are same as those given in section II. By fitting $R_{42} \equiv E_{4_1^+}/E_{2_1^+} = 2$ for vibrational case, the parameters used to produce the vibrational spectra were obtained, and presented in Table 2. In the SDPSM, the full shell model space was truncated to the SD-pair subspace. The investigation on the validity of the SD-pair truncation in Ref.[17-19] show that the SD-pair truncation can not produce the rotational spectra. But Dr. Zhao's work[12] and our previous work[7] show that if a reasonable Hamiltonian and a suitable collective SD-pair structure were considered, the rotational behaviors can be produced very well. It is found that with $2\kappa_\pi = 2\kappa_\nu = \kappa = 0.2\text{MeV}/r_0^4$, the similar results to the $SU(3)_\pi \times SU(3)_\nu$ limit of the IBM can be reproduced, in which the typical energy ratios $E_{4_1^+}/E_{2_1^+}$ and $E_{6_1^+}/E_{2_1^+}$ are 3.33 and 6.96, close to the IBM result 3.33 and 7, respectively. The detailed discussion can be found in Refs.[7,12]

Table 2. The parameters used to produce the vibrational, rotational spectra. G_σ is in unit of MeV, κ_σ and κ are in unit of MeV/r_0^4.

limit	G_π	G_ν)	κ_π	κ_ν	κ
vibration	0.5	0.5	0	0	0.01
rotation	0	0	0.1	0.1	0.2

To identify shape phase transitions and determine the corresponding patterns, Iachello *et al* initialed a study on effective order parameters, which should display different critical behaviors for the phase transitions with different order. Specifically, the quantities related with isomer shifts, defined as $v_2 = (< 0_2^+|\hat{n}_d|0_2^+ > - < 0_1^+|\hat{n}_d|0_1^+ >)/N$ and $v_2' = (< 2_1^+|\hat{n}_d|2_1^+ > - < 0_1^+|\hat{n}_d|0_1^+ >)/N$, were proposed as effective-order parameters in.[20] Consequently, some other quantities, such as the $B(E2)$ ratios $K_1 = B(E2; 4_1^+ \to 2_1^+)/B(E2; 2_1^+ \to 0_1^+)$ and $K_2 = B(E2; 0_2^+ \to 2_1^+)/B(E2; 2_1^+ \to 0_1^+)$[21] as well as the energy ratio $R_{60} = E_{6_1^+}/E_{0_2^+}$ were also suggested as the effective order parameters to identify phase transitions and the corresponding orders. Therefore, to study the shape phase transition in the SD-pair fermion model space, v_2, v_2', in which the d-boson number operator $\hat{n}_d$ is replaced by D-pair number operator $\hat{N}_D$ in the SDPSM, K_1, K_2 and R_{60} will be studied in this paper. Because the importance of $R_{42} = E_{4_1^+}/E_{2_1^+}$ in determining the limiting cases and shape phase transitions,[22] R_{42} is also presented.

Energy ratios R_{42} and R_{60} against control parameter α are shown in Fig. 3. Figure 3a shows that the energy ratio R_{42} is 2 (when $\alpha = 0$) and 3.3 (when $\alpha = 1$), which are typical values of vibrational and rotational spectra, respectively, in the IBM.[23] It is also shown that the rapid change occurs when $0.3 \le \alpha \le 0.6$, which indicates that the phase transition occurs within this region.

The energy ratio R_{60} given in Fig. 3b shows that similar behavior to that of the IBM for finite number of boson N_B is reproduced. It exhibits a modest peak followed by a sharp decrease across the phase transition, a typical signature of the 1st-order quantum phase transition.[24]

The SDPSM results of v_2, v_2', K_1 and K_2 are given in Fig. 4 and Fig. 5. The effective charges were fixed with $e_\pi = 3e_\nu = 1.5e$. As argued in,[20] v_2, v_2' should have wiggling behaviors in the region of the critical point due to the switching of the two coexisting phases for the first order phase transition. Indeed, the obvious wiggling behaviors shown by v_2, v_2' in Fig. 4

Table 3. Energy and B(E2) ratios at vibrational, rotational limit, and $X(5)$-like critical point calculated in the SDPSM.

limit	$\dfrac{E_{4_1^+}}{E_{2_1^+}}$	$\dfrac{E_{6_1^+}}{E_{2_1^+}}$	$\dfrac{E_{6_1^+}}{E_{0_2^+}}$	$\dfrac{4_1^+\to 2_1^+}{2_1^+\to 0_1^+}$	$\dfrac{6_1^+\to 4_1^+}{2_1^+\to 0_1^+}$
vibrational limit	1.99	2.97	1.47	1.49	1.48
$X(5)$-like point	2.91	5.60	1.05	1.38	1.38
rotational limit	3.33	6.96	0.46	1.34	1.32
	$\dfrac{E_{0_2^+}}{E_{2_1^+}}$	$\dfrac{E_{2^+}-E_{0_2^+}}{E_{2_1^+}}$	$\dfrac{E_{4^+}-E_{0_2^+}}{E_{2_1^+}}$	$\dfrac{2^+\to 0_2^+}{2_1^+\to 0_1^+}$	$\dfrac{4^+\to 2^+}{2_1^+\to 0_1^+}$
$X(5)$-like point (0_2^+ band)	5.32	2.30	5.33	0.37	0.43

further confirm the transition is first order. The results of $B(E2)$ ratio K_1 is also consistent with those of other effective quantities.[20,21] The critical behavior of K_2 seems to deviate from the character of the first order phase transition.

In the IBM, the critical point symmetry[25] between $U(5)$ and $SU(3)$ is $X(5)$. Since the shape phase transition between vibrational and rotational limit can be reproduced in the SDPSM, it is interesting to see if the properties of the $X(5)$-like symmetry also occurs within the SDPSM. We found that there is indeed a signature with $\alpha = 0.54$ in the SDPSM similar to that of the $X(5)$ in the IBM. A few typical values are given in Table 3, from which one can see that typical feature of the $X(5)$ symmetry stated in Ref.[24,26] indeed occurs in the SDPSM. For example, R_{42}, R_{60} and $E_{0_2^+}/E_{2_1^+}$ is 2.91, 1.05 and 5.32 in the SDPSM calculation, close to the IBM results 2.91, 1.0 and 5.67, respectively.

4. Summary

In summary, the vibrational, γ-unstable, rotational and triaxial spectra corresponding to the $U(5)$, $SO(6)$, $SU(3)$ and $SU^*(3)$ limiting cases in the IBM can indeed be reproduced in the nucleon-pair shell model truncated to a SD subspace (SDPSM). The nuclear shape phase transitional patterns of vibration-rotation and vibration-γ-soft are indeed similar to the corresponding results obtained from the IBM previously. The signatures of the critical point symmetry in the SD-pair shell model are also close to those shown in the IBM. The analysis not only shows that the IBM has a sound shell-model foundation, but also confirms that the truncation scheme adopted in the SD-pair shell model seems reasonable as long as the Hamil-

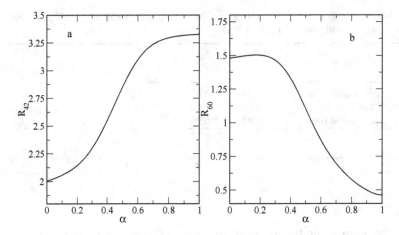

Fig. 3.   Energy ratios R_{42} and R_{60} vs α for the vibration-rotation transition.

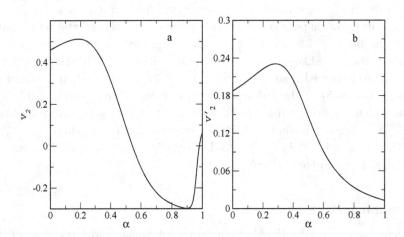

Fig. 4. v_2 and v_2' vs α in the vibration-rotation transition.

tonian is reasonably chosen even when realistic single-particle energies are taken into consideration. The results suggest the value of further analysis using the SDPSM to see whether or not shape phase transitions could be described in terms of the nucleon degrees of freedom. Refs.[25] show that there are critical point symmetries such as X_5 and E_5 predicted from the Bohr Hamiltonian. It should be a interesting exercise to describe such criti-

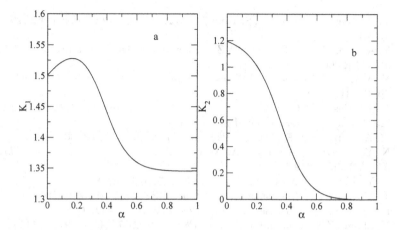

Fig. 5. B(E2) ratios vs α in the vibration-rotation transition.

cal point phenomena within the framework of the SDPSM with its fermion foundation.

Acknowledgments

Support from the Natural Science Foundation of China (11075080), and the LSU–LNNU joint research program (C164063) is acknowledged.

References

1. J. Q. Chen, *Nucl. Phys. A* **562**, 218 (1993).
2. J. Q. Chen, *Nucl. Phys. A* **626**, 686 (1997).
3. Y. M. Zhao, N. Yoshinaga, S. Yamaji, J. Q. Chen and A. Arima, *Phys. Rev. C* **62**, 014304 (2000).
4. K. Allaart, E. Boeker, G. Bonsignori, M. Savoia and Y. K. Gambhir, *Phys. Rep.* **169**, 209 (1988).
5. K. T. Hecht, J. B. McGrory and J. P. Draayer, *Nucl. Phys. A* **197**, 369 (1972).
6. C. L. Wu, D. H. Feng, X. G. Chen, J. Q. Chen and M. Guidry, *Phys. Rev. C* **36**, 1157 (1987).
7. Y. A. Luo, F. Pan, C. Bahri, and J. P. Draayer, *Phys. Rev. C* **71** (2005) 044304.
8. L. Li, Y. A. Luo, T. Wang, F. Pan and J. P. Draayer, *J. Phys. G: Nucl. Part. Phys.* **36**, 125107 (2009).
9. A. E. L. Dieperink and R. Bijker, *Phys. Lett. B* **116** (1982) 77.
10. M. A. Caprio and F. Iachello, *Ann. Phys.* **318** (2005) 454.
11. A. Sevrin K. Heyde and J. Jolie, *Phys. Rev. C* **36** (1987) 2621.

140

12. Y. M. Zhao, N. Yoshinaga, S. Yamaji, and A. Arima, *Phys. Rev. C* **62** (2000) 014316.
13. Y. A. Luo, F. Pan, T. Wang, P. Z. Ning and J. P. Draayer, *Phys. Rev. C* **73**, 044323 (2006).
14. Y. A. Luo, Y. Zhang, X. F. Meng, F. Pan and J. P. Draayer, *Phys. Rev. C* **80**, 014311 (2009).
15. J. M. Arias, J. E. García-Ramos and J. Dukelsky, *Phys. Rev. Lett.* **93**, 212501 (2004) and the reference cited in this paper.
16. M. A. Caprio and F. Iachello, *Phys. Rev. Lett.* **93**, 242502(2004).
17. N. Yoshinaga, T. Mizusaki, A. Arima and Y. D. Devi, *Prog. Theor. Phys. Suppl.* **125**, 65 (1996).
18. N. Yoshinaga, D. M. Brink, *Nucl. Phys. A* **515**, 1 (1990).
19. T. Mizusaki and T. Otsuka, *Prog. Theor. Phys. Suppl.* **125**, 97 (1996).
20. F. Iachello and N. V. Zamfir, *Phys. Rev. Lett.* **92**, 212501 (2004).
21. Y. Zhang, Z. F. Hou and Y. X. Liu, *Phys. Rev. C* **76**, 011305(R) (2007).
22. V. Werner, P. von Brentano, R. F. Casten and J. Jolie, *Phys. Lett. B* **527**, 55 (2002).
23. F. Iachello and A. Arima, *The Interacting Boson Model*, Cambridge University Press, Cambridge New York, 1987.
24. D. Bonatsos, E. A. McCutchan, R. F. Casten and R. J. Casperson, *Phys. Rev. Lett.* **100**, 142501 (2008).
25. F. Iachello, *Phys. Rev. Lett.* **87**, 052502 (2001).
26. R. M. Clark *et al.*, *Phys. Rev. C* **68**, 037301 (2003).

CROSS SECTION DISTRIBUTIONS OF FRAGMENTS IN THE CALCIUM ISOTOPES PROJECTILE FRAGMENTATION AT THE INTERMEDIATE ENERGY

C. W. MA*, Y. F. ZHANG, and H. L. WEI

Department of Physics, Henan Normal University,
Xinxiang, Henan Province 453007, China
*E-mail: machunwang@126.com

The cross sections of fragments (σ_f) produced in the 100 A MeV even $^{36-52}$Ca projectile fragmentation reactions are evaluated in the framework of the statistical abrasion-ablation model. The distributions of σ_f are compared and the similarities of σ_f distributions are investigated.

Keywords: Heavy-ion reactions; the statistical abrasion-ablation model; Isospin phenomenon.

1. Introduction

The constructions of the third generation of radioactive nuclear beam facilities around the world have greatly stimulated the research in phenomena in heavy-ion collisions induced by the very neutron-rich nucleus, of whom has big isospin freedom.[1] Due to the difficulties to study the density distribution of neutron in the very neutron-rich nucleus directly, phenomena related to density distributions of neutrons and protons are explored to extract information of neutron indirectly.[2-6,8-10,14]

2. Model and results

The SAA model was developed to describe heavy-ion collisions both at high and intermediate energies.[11-13] This model takes independent N-N collisions for participants in an overlap zone of the two colliding nuclei and determines the distributions of abraded neutrons and protons. The details of SAA model can be found in Refs.[3,6,8-13]

Using the SAA model, the cross sections of fragments produced in the 100 A MeV even $^{36-52}$Ca+^{12}C are calculated and plotted in Fig. 1. Fitting

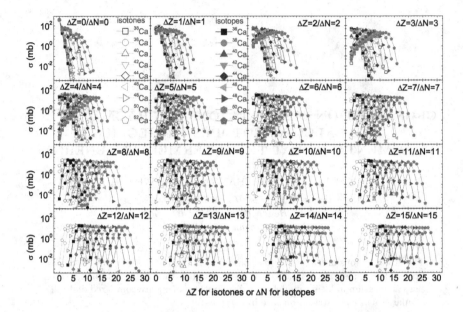

Fig. 1. (Color online) Cross sections of fragments produced in the 100 A MeV even $^{36-52}$Ca+^{12}C using the SAA model. For isotopes, the x axes represent the removed neutron numbers from the projectile($\Delta N = N_p - N_f$); for isotones, the x axes represent the removed proton numbers from the projectile($\Delta Z = Z_p - Z_f$).

these distributions using a Gaussian function, peak positions are obtained and plotted in Fig. 2.

The peak positions of these isotopes or isotones increase linearly except the isotopes or isotones with small ΔZ or ΔN. The peak positions of isotopes and isotones of ^{40}Ca have very little difference, while peak positions of isotopes and isotones produced of the neutron-proton asymmetric projectile nuclei become bigger as the n/p ratios of projectile nuclei increase.

In Refs.,[8,9] it is discussed that the neutron-removal cross section for isotopes could be used as an observable to extract neutron-skin thickness of neutron-rich nucleus. Fragments isotopes and isotones with small ΔZ or ΔN are mostly produced in the peripheral collisions[9,10] and the distributions of cross sections could not be fitted using the Gaussian function. Isotopes with $\Delta Z \leq 3$ and isotones with $\Delta N \leq 3$ are plotted in Fig. 3 and Fig. 4. For isotopes, the distributions of the Z=17-20 fragments are very similar for the $^{36-40}$Ca projectile. For other Ca projectile, the distributions of isotope fragments show big difference. For isotones, the distributions of

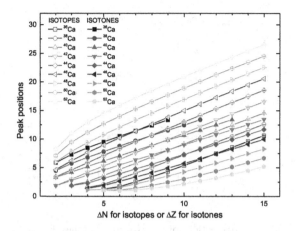

Fig. 2. (Color online) Peak positions of isotopes and isotones in Fig. 1. The open symbols are for isotopes and the filled ones are for isotones.

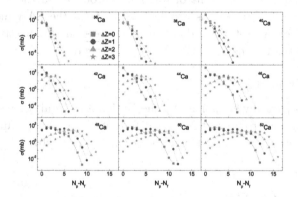

Fig. 3. (Color online) Cross sections of the ΔZ=0-3 isotopes. The x axes represent the removed neutrons from the projectile.

the ΔN=0-3 fragments are very similar for the neutron-rich $^{46-52}$Ca projectile. For other Ca projectile, the distributions of isotope fragments show big difference. It's easy to see that for proton-rich and symmetric nuclei the isotopic distributions of fragments are very similar, and for neutron-rich nuclei the isotonic distributions of fragments are very similar. The ratios of isotopic cross sections have been used to extract the symmetry energy using the isoscaling method.[14] The ratios of isotonic cross section may be used to extract symmetry energy but the method will be more difficult than the isotopic method.

144

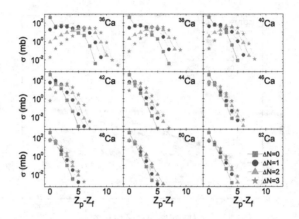

Fig. 4. (Color online) Cross sections of the ΔN=0-3 isotones. For isotones, the x axes represent the removed protons from the projectile.

In summary, cross sections of fragments produced in the 100 A MeV even $^{36-52}$Ca $+^{12}$C reactions are calculated using the SAA model. The distributions of the isotopic and isotonic cross sections are investigated and similarity of the fragment distributions are compared.

Acknowledgments

This work is partially supported by the National Natural Science Foundation of China under contract No. 10905017, and the Youth Foundation in Science of Henan Normal University under grant 2009qk07.

References

1. Bao-An Li *et al.*, *Phys. Rep.* **464**, 113 (2008).
2. X. Y. Sun *et al.*, *Phys. Lett. B* **687**, 396 (2010).
3. C. W. Ma *et al.*, *Phys. Rev. C* **79**, 034606 (2009).
4. C. W. Ma *et al.*, *IJMPE* **17**, 1669 (2008).
5. Y. G. Ma *et al.*, *Phys. Rev. C* **60**, 024607 (1999).
6. C. W. Ma *et al.*, *Chin. Phys. B*, **17**, 1216 (2008).
7. H. S. Xu *et al.*, *Phys. Rev. Lett.* **85**, 716 (2000).
8. D. Q. Fang *et al.*, *Phys. Rev. C* **82**, 047603 (2010).
9. Chun Wang Ma *et al.*, *Phys. Rev. C* **82**, 057602 (2010).
10. Chun-Wang Ma *et al.*, *J. Phys. G: Nucl. Part. Phys.* **37**, 015104 (2010).
11. T. Brohm, and K. -H. Schmidt, *Nucl. Phys A* **569**, 821 (1994).
12. J. J. Gaimard, and K. -H. Schmidt, *Nucl. Phys. A* **531**, 709 (1991).
13. D. Q. Fang *et al.*, *Phys. Rev. C* **61** 044610 (2000).
14. H. S. Xu *et al.*, Phys. Rev. Lett. **85**, 716 (2000).

SYSTEMATIC STUDY OF SPIN ASSIGNMENT AND DYNAMIC MOMENT OF INERTIA OF HIGH-J INTRUDER BAND IN ^{111,113}In

K. Y. MA, D. YANG, J. B. LU*, H. D. WANG, Y. Z. LIU, G. Y. LIU, L. LI, Y. J. MA, S. YANG, S. LIU

Department of Physics, Jilin University,
Changchun, 130012, China
*E-mail: ljb@jlu.edu.cn

L. H. ZHU, X. G. WU, G. S. LI, C. Y. HE, X. Q. LI

China Institute of Atomic Energy,
Beijing, 102413, China

L. L. WANG

Southwest University of Science and Technology,
Sichuan, 621010, China

The method of excitation energy systematics has been used to help the spin assignment of the $\pi h_{11/2}$ intruder band in ^{111,113}In. $I_0=23/2$ and $I_0 = 15/2$ are assigned to the lowest observed state of the intruder band in ^{111,113}In. The dynamic moment of inertia and the frequency of the $h_{11/2}$ neutrons' aligning of the $\pi h_{11/2}$ intruder band is compared with in the neighboring nuclei. The shift of the crossing frequency is mainly caused by the p-n interaction and the interaction becomes weaker with the number of the neutron increasing.

Keywords: Excitation energy; dynamic moment of inertia; intruder band.

1. Introduction

Nuclei near the $Z=50$ shell closure have been found to exhibit collective structures coexisting with the single particle states. The deformed states in Sn nuclei are due to the proton two-particle-two-hole (2p-2h) excitation across the $Z=50$ shell gap, the underlying configuration being $\pi(g_{7/2})^2(g_{9/2})^{-2}$ [1]. In the odd mass Sb [2-10] isotopes, the low-lying states represent the single-particle structure, which indicates the valence proton orbitals $g_{7/2}$, $d_{5/2}$, and $h_{11/2}$ are coupled to the spherical Sn core states, and the high-spin states are dominated by the intruder band. The $\triangle I = 1$ collective band is observed and interpreted as two-particle-one-hole (2p-1h) excitation based on the $\pi(g_{7/2})^2(g_{9/2})^{-1}$ configuration [3-8]. Furthermore, the $\triangle I = 2$ decoupled rotational band has been identified involving the valence nucleons occupying orbitals $g_{7/2}$, $d_{5/2}$, and $h_{11/2}$ coupled to the 2p-2h deformed core Sn states [2-9]. In addition, a unique phenomenon called smooth band termination has been discovered in some of these intruder bands in Sb [9,10], Te [11] and I isotopes [12,13]. The richness of information on the coexisting spherical, nearly spherical and deformed structures are observed above the $Z \geqslant 50$ nuclei, which attracts the interest in the investigation of the structure of the nuclei below the Z=50 gap. Several experiments have been performed on the In isotopes, recently [14-19]. However, the decoupled intruder bands are only found in two nuclei of $^{111,113}In$ [15-17]. These bands are not linked to states in the known level schemes, so the spin-parity is assigned tentatively. In this work, the method of excitation energy systematics is used to help the spin assignment.

2. Excitation energy systematics of the intruder band in ^{111,113}In

The method of excitation energy systematics has been used to help the spin assignment of the $\pi(h_{11/2})_p \otimes \upsilon(i_{13/2})_n$ bands in $A \sim 160$ and $\pi(h_{11/2})_p \otimes \upsilon(h_{11/2})_n$ bands in $A \sim 130$ nuclei [20,21]. The main idea of this method is that in the bands of the same configuration, the excitation energy of the levels with the same spin of a chain of isotopes (isotones) varies with neutron (proton) number in a smooth way. The most intense decoupled band in ^{111,113}In [15-17] involves 1p-1h proton excitation with the $\pi h_{11/2}(g_{9/2})^{-2}$ configuration, in ^{113,115}Sb [6,8] involves a 2p-2h proton excitation Sn core coupled with the valence proton in $h_{11/2}$ orbital, i.e., the $\pi h_{11/2}(g_{7/2})^2(g_{9/2})^{-2}$ configuration, and in ^{115,117}I [22,23] and ^{117,119}Cs [24,25] is of the $h_{11/2}[550]1/2^-$ configuration. The $h_{11/2}$ decoupled band in In isotopes has a smaller deformation (β_2=0.18 for ^{111}In [26] comparing with β_2=0.32 for ^{113}Sb [6]).

In ^{111}In, I_0=23/2 was assigned to the lowest observed state of the decoupled band. However, the two experiments of ^{113}In [16,17] give deferent spin assignment to the decoupled band and the deference is $2\hbar$. The $h_{11/2}$ decoupled band in ^{113,115}Sb and ^{115,117}I has been connected with the low-lying levels and so the spin-parity quantum number has been firmly assigned. The Fig.1 shows the energy systematics of the $h_{11/2}$ decoupled band in N=62 and N=64 odd-A isotones. The level with $I = 23/2$ is used as zero-energy reference. It shows that $I_0=23/2$ for ^{111}In and $I_0=15/2$ for ^{113}In fit to the smooth trend, so the $I_0=23/2$ for ^{111}In and $I_0=15/2$ for ^{113}In are the reasonable assignment. The result agrees with the assignment in [15,17], and disapproves the assignment in [16].

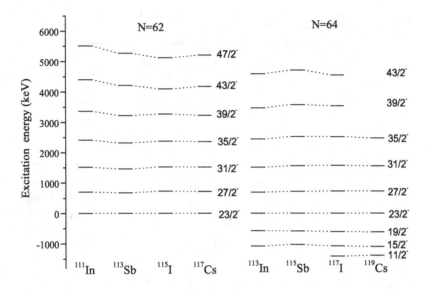

Fig. 1.   Excitation energy systematics of the $N=62$, 64 isotones.

3. Systematic study of the the dynamic moment of inertia of the $N=62$, 64 isotones

It is instructive to study the properties of the dynamic moment of inertia of the $h_{11/2}$ intruder band in the neighboring nuclei. Fig. 2 shows the systematic study of the dynamic moment of inertia of the intruder bands in the $N=62$ and $N=64$ odd-A isotones. It shows that the dynamic moment of inertia is nearly the same in a chain of isotones and the line becomes much more gradual with the number of neutrons increasing. It suggests that the dynamic moment of inertia of the $h_{11/2}$ decoupled band depends on the orbital occupied by the $h_{11/2}$ neutron. As shown in Table 1, the delay of the first neutron alignment is systematically delayed comparing with the even-even Sn. For these nuclei, the aligning $h_{11/2}$ neutrons are in nearly the same orbitals as the proton. It suggests the shift of the crossing frequency is mainly caused by the p-n interaction. The quadrupole type of p-n interaction has been used to calculate this shift in the crossing frequency in the earlier work [6]. The crossing frequencies of I and Cs are both lower than that of In and Sb. It suggests that the p-n interaction becomes weaker as the number of neutrons increases.

Table 1. $vh_{11/2}$ crossing frequencies (MeV) for the $N=62$, 64 odd-A isotones (part from [16]).

Neutrons	Sn	In	Sb	I	Cs
62	0.37	0.45	0.46	0.42	0.42
64	0.41	> 0.55	> 0.5	0.52	> 0.42

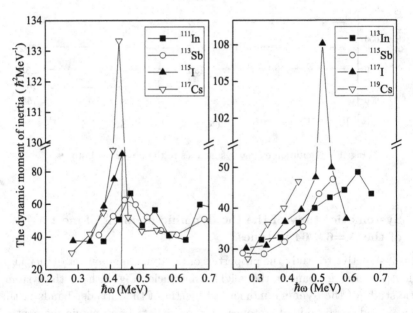

Fig. 2. The dynamic moment of inertia in the $N=62$ and $N=64$ odd-A isotones.

4. Summary

In this work, the method of excitation energy systematics has been used to help the spin assignment of the $h_{11/2}$ intruder band of the [111,113] In. The results agrees with the assignment in [15,17], and disapproves the assignment in [16]. The dynamic moment of inertia and the frequency of the $h_{11/2}$ neutrons' aligning of the $h_{11/2}$ intruder band is comparing in the neighboring nuclei. It is suggested the shift of the crossing frequency is mainly caused by the p-n interaction and the interaction becomes weaker with the number of the neutron increasing.

Acknowledgments

We would like to thank the operating staff the HI-13Tandem Accelerator in the China Institute of Atomic Energy for help.This work is supported by the National Natural Science Foundation of China under Grant No 10105003, No 11075064, No 11075214 and No 10927507 the Specialized Research Fund for the Doctoral Programme of Higher Education of China under Grant No 20050183008, and the National Basic Research Programme of China under Grant No 2007CB815005 and No 2007CB815000.

References

1. J. Bron et al., Nucl. Phys. A **318**, 335 (1979).
2. T. Ishii et al., Phys. Rev. C **49**, 2982 (1994).
3. D. R. LaFosse et al., Phys. Rev. C **50**, 1819 (1994).
4. J. Shergur et al., Phys. Rev. C **71**, 064323 (2005).
5. R. E. Shroy et al., Phys. Rev. C **19**, 1324 (1979).
6. V. P. Janzen et al., Phys. Rev. Lett. **70**, 1065 (1993).
7. C. -B. Moon et al., Phys. Rev. C **58**, 1833 (1998).
8. R. S. Chakrawarthy et al., Phys. Rev. C **54**, 2319 (1996).
9. H. Schnare et al., Phys. Rev. C **54**, 1598 (1996).
10. V. P. Janzen et al., Phys. Rev. L **72**, 1060 (1994).
11. I. Thorsland et al., Phys. Rev. C **52**, R2839 (1995).
12. E. S. Paul et al., Phys. Rev. C **50**, 741 (1994).
13. E. S. Paul et al., Phys. Rev. C **59**, 1984 (1999).
14. J. Kownacki et al., Nucl. Phys. A **627**, 239 (1997).
15. P. Vaska et al., Phys. Rev. C **57**, 1634 (1998).
16. R. S. Chakrawarthy et al., Phys. Rev. C **55**, 155 (1997).
17. S. Naguleswaran et al., Phys. Rev. C **72**, 044304 (2005).
18. S. Naguleswaran et al., Z. Phys. A **359**, 235 (1997).
19. R. Lucas et al., Eur. Phys. J. A **15**, 315 (2002).
20. Yunzuo Liu et al., Phys. Rev. C **52**, 2514 (1995).
21. Yunzuo Liu et al., Phys. Rev. C **54**, 719 (1996).
22. E. S. Paul et al., J. Phys. G **18**, 837 (1992).
23. C. -B. Moon et al., Eur. Phys. J. A **5**, 13 (1999).
24. J. F. Smith et al., Phys. Rev. C **63**, 024319 (2001).
25. F. Liden et al., Nucl. Phys. A **550**, 365 (1992).
26. S. M. Mullins et al., Phys. Lett. B **3184**, 592 (1993).

SIGNALS OF DIPROTON EMISSION FROM THE THREE-BODY BREAKUP CHANNEL OF ^{23}AL AND ^{22}MG[*]

MA YU-GANG[†1], FANG DE-QING[1], SUN XIA-YIN[1,2], ZHOU PEI[1,2], CAI XiANG-ZHOU[1], CHEN JIN-GEN[1], GUO WEI[1], TIAN WEN-DONG[1], WANG HONG-WEI[1], ZHANG GUO-QIANG[1,2], CAO XI-GUANG[1], FU YAO[1], HU ZhENG-GUO[3], WANG JIAN-SONG[3], WANG MENG[3], TOGANO Y[4], AOI N[4], BABA H[4], HONDA T[5], OKADA K[5], HARA Y[5], IEKI K[5], ISHIBASHI Y[6], ITOU Y[6], IWASA N[7], KANNO S[4], KAWABATA T[8], KIMURA H[9], KONDO Y[4], KURITA K[5], KUROKAWA M[4], MORIGUCHI T[6], MURAKAMI H[4], OISHI H[6], OTA S[8], OZAWA A[6], SAKURAI H[4], SHIMOURA S[8], SHIODA R[5], TAKESHITA E[4], TAKEUCHI S[4], YAMADA K[4], YAMADA Y[5], YASUDA Y[6], YONEDA K[4] and MOTOBAYASHI T[4]

[1]Shanghai Institute of Applied Physics, Chinese Academy of Sciences, Shanghai 201800, China;
[2]Graduate School of the Chinese Academy of Sciences, Beijing 100039, China;
[3]Institute of Modern Physics, Chinese Academy of Sciences, Lanzhou 730000, China;
[4]Institute of Physical and Chemical Research (RIKEN), Wako, Saitama 351-0198, Japan;
[5]Department of Physics, Rikkyo University, Japan;
[6]Institute of Physics, University of Tsukuba, Ibaraki, Japan;
[7]Department of Physics, Tohoku University, Japan;
[8]Center for Nuclear Study (CNS), University of Tokyo, Japan;
[‡9]Department of Physics, University of Tokyo, Japan

Two-proton relative momentum from the break-up channels ^{23}Al $\rightarrow$ p + p + ^{21}Na and ^{22}Mg $\rightarrow$ p + p + ^{20}Ne at an energy of 72 AMeV has been measured together with two-proton opening angles at the projectile fragment separator beamline (RIPS) in the RIKEN Ring Cyclotron Facility. The results demonstrate that there exists diproton emission component from single-step ^{2}He for highly excited nuclei of ^{23}Al and ^{22}Mg.

1. Introduction

Studies on exotic nuclear reaction and structure are the current focus in low-intermediate energy nuclear physics. Many new phenomena were observed,

---

[*]This work is supported by National Natural Science Foundation of China under contract Nos 10775167, 10775168, 10979074, 10405032, 10747163, 10805067 and 10605036, Major State Basic Research Development Program in China under contract No. 2007CB815004, and the Shanghai Development Foundation for Science and Technology under contract No. 09JC1416800.
[†]Invited plenary speaker and corresponding author. Email: ygma@sinap.ac.cn.

such as neutron-halo, pygmy dipole resonance and new magic-number for nuclei far from the β-stability line. With the advent of the advanced radioactive ion beam and nuclear detection technique, more precise experiments on nuclear structure and reaction of extremely neutron- or proton-rich nuclei can be performed.

For the proton-rich nuclei, the proton decay mechanism is complicated, especially for two-proton (2p) radioactivity [1]. Usually, there are three possible ways for two proton emission: (i) two-body sequential emission in a short time; (ii) three-body simultaneously democratic emission; and (iii) ^{2}He cluster emission and then breakup into two protons. The later two protons in a ^{2}He cluster are basically constrained by the pair correlation in a quasi-bound s-singlet, i.e., $1s$ configuration. The Coulomb barrier can guarantee the existence of such a quasi-bound state for an instant. After penetration of the barrier, the two protons will be separated. The experimental search for the 2p emitter started very early [2,3], and some other experiments and modern theories are still required to comprehensively understand the decay mechanism [4-6]. The current progress of experiment and theory on 2p radioactivity can be found in a recent review paper [7].

Full decay channels from cold or low-excited nuclei could be reconstructed by the advanced detector arrays. For instance, one can identify the decay channel $(A-2, Z-2) + p + p$ from a proton-rich nucleus (A, Z) by the Si-strip and other ΔE multi-detectors combination. In this decay channel, relative momentum and opening angle between two protons can be reconstructed. Different from the p-p correlation of hot nuclei, here p-p correlation can mostly give the information of internal proton-proton binding or decay mode in a specific channel. As we mentioned before, there are three major types of proton emission mechanism. Of which, ^{2}He cluster emission is of the most interesting. In this case, a strong correlation of p-p relative momentum around 20 MeV/c will appear together with the small opening angle between two protons in the rest frame of three body decay.

For the proton-rich nucleus ^{23}Al, it has attracted a lot of interest in recent years for some reasons. A rather small proton separation energy of 0.122 MeV [8], makes it a candidate for a proton-halo nucleus. "An abnormal increase" in the reaction cross section of ^{23}Al has been observed in our previous experiment performed at HIRFL-RIBLL beamline, Lanzhou [9] as well as in our later experiment performed at RIKEN-RIPS beamline. However, the fragment momentum distribution of ^{22}Mg is normal in our RIPS experiment [10]. With the help of the few-body Glauber model, it shows that the valence proton of ^{23}Al is dominant by d-wave [10]. In addition, the spin and parity of the ^{23}Al ground state was found to be $J^{\pi}= 5/2^{+}$ by recent β -NMR [11] and β -decay measurements [12], which is in agreement with the measurement to its mirror nucleus ^{23}Ne. The mass excess of the proton-rich nucleus ^{23}Al has been recently measured with the JYFLTRAP Penning trap setup [13]. ^{23}Al may also play a crucial role in solving the depletion of the NeNa cycle in ONe novae. The key astrophysical reactions related to ^{23}Al and ^{22}Mg (p, γ)^{23}Al are discussed in [14] and references therein.

On the other hand, the structure of proton-rich nucleus ^{22}Mg has received great interest in recent years because of its importance in determining the astrophysical reaction rates of ^{21}Na (p, γ) ^{22}Mg and ^{18}Ne(α,p)^{21}Na reactions in explosive stellar scenarios [15,16] . The excited states in ^{22}Mg have been investigated via various reactions, such as ^{24}Mg(p, t)^{22}Mg, ^{24}Mg(α, ^{6}He) and ^{12}C(^{16}O, ^{6}He) reactions, the ^{25}Mg(^{3}He, ^{6}He) reaction, the ^{20}Ne(^{3}He, n) reaction as well as the ^{20}Ne(^{3}He, n γ) reaction etc. By knowing the location of excitation states in ^{22}Mg, the resonance property of states just above the proton threshold has been studied by the direct ^{21}Na(p, γ)^{22}Mg measurements [17,18]. The astrophysical implication of the ^{21}Na(p, γ)^{22}Mg reaction has been discussed on a firm experimental result [19].

In the present paper, an exclusive measurement has been analyzed so that three body decay channels from ^{23}Al and ^{22}Mg, respectively, are selected. Excitation energy spectra are reconstructed by the invariant mass method. By investigating the relative momentum and opening angle between two protons in different excitation energy windows, we search for the diproton emission process. Our results show there exists such kind of emission mechanism in high excitation energy region.

2. Description of experiment

The experiment was performed at the RIKEN-RIPS beamline in the RIKEN Ring Cyclotron Facility. The secondary ^{23}Al and ^{22}Mg beams with incident energy of 72 AMeV were generated by projectile fragmentation of 135A MeV

^{28}Si primary beam on ^{9}Be production target at F0 chamber and then transported to a ^{12}C reaction target at F2 chamber. In the dispersive focus plane F1, an Al wedge-shaped degrader (central thickness: 583.1 mg/cm^2, angle: 6 mrad) was installed. A delay-line readout parallel plate avalanche counter (PPAC) was placed to measure the beam position. Then the secondary beam was directed onto the achromatic focus F2. Two delay-line readout PPACs were installed at F3 to determine the beam position, angle and also were used to extract the position on the target that the beams hit. A silicon detector F3SSD was installed after the two PPACs to measure the energy loss (ΔE) of the secondary beams. An ultra fast plastic scintillator (F3, 0.5mm thick) was placed before a carbon reaction target (355.5 mg/cm^2 thick) to measure the time-of-flight (TOF) from the plastic scintillator at F1 and F2. The experimental setup before the reaction target was similar to the previous experiment [14].

In the downstream of the reaction target, there were five layers of silicon detector. The first two layers of Si-strip (5mm width) detectors located around 50cm downstream of the target were used to measure the emitted angle of the fragment and protons, and the other three layers of element Si detectors were used as the ΔE-E detectors for the fragment. Each Si-strip layer consists of 5×5 matrix without detectors in the four corners. The plastic hodoscopes located around 3m downstream of the target were used as ΔE, E and TOF detectors for protons. Most of the protons stopped before the third layer.

Several radioactive beams are delivered from RIPS beamline. The particle identification (PID) of ^{23}Al and ^{22}Mg before the reaction target was done by means of Bρ-ΔE-TOF method. After the reaction target, the heavy fragments were identified by five layers silicon detectors combination through the ΔE -E technique. Both the emitted angle and energy loss can be obtained for the fragments. The total energy of heavy fragments can be obtained by the sum over the energy loss of the five layers silicon detectors. From this setup, less than 5 MeV/c of the relative momentum resolution for protons at the typical energy of 65 MeV could be achieved.

3. Preliminary results and discussions

Clear particle identification can be obtained and the exclusive measurement from the break-up of radioactive beam can be realized. In our present analysis,

154

the reaction channel of ^{21}Na + p + p was picked out and the excitation energy of ^{23}Al was reconstructed by the difference between the invariant mass of three-body decay channel and its mother-nucleus mass in the ground state,

$$E^* = W - M = \sqrt{(\sum E_i)^2 - (\sum \vec{P}_i c)^2} - (M_{^{23}Al}). \quad (1)$$

Generally, diproton emission process from the ground state is rare. This emission is also called ground-state two-proton radioactivity which was observed in a few nuclei [20]. Two-proton radioactivity was predicted to occur for even-Z nuclei, for which, due to the pairing force, one proton emission is energetically forbidden, whereas two-proton emission is allowed. As two-proton emission is a process, which is essentially governed by the Coulomb and centrifugal barriers, it was quickly seen that this condition can be fulfilled only for nuclei with a reasonably high Coulomb barrier. However, diproton emission from ^{2}He becomes more favorable in the excited states for proton-rich nuclei.

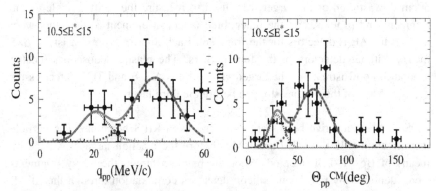

Fig. 1. Left panel: relative momentum spectrum of two protons for ^{23}Al's decay into two protons and ^{21}Na in 10.5 ≤E*≤15 MeV. Right panel: opening angular distribution between two protons. The short-dotted line and long-dashed line represent two separate Gaussian fits and the solid line represents the sum of two Gaussian fits.

In our present experiment, we investigate the relative momentum spectrum (q_{pp}) and opening angle (θ_{pp}) of the two protons for ^{23}Al in different excitation energy windows. In higher energy region, there exists an obvious component which peaks at small relative momentum and opening angle. Fig. 1 shows the preliminary result of the above two distributions in the rest frame of three-body decay system in the excitation energy window (10.5 ≤ E* ≤15 MeV) for ^{23}Al.

A peak in relative momentum q_{pp} around 20 MeV/c is clearly observed which corresponds to a peak in smaller opening angle θ_{pp} around 30°. Besides, another peak is observed around q_{pp} ~ 40 MeV/c which corresponds to the component of larger θ_{pp}. The peak at 20 MeV/c of q_{pp} and small opening angle is consistent with the diproton emission mechanism. In contrary, the peak at larger q_{pp} and θ_{pp} may correspond to sequential proton decay or three body democratic decay.

In order to quantitatively extract the diproton emission probability in above excitation energy window of ^{23}Al, two Gaussian fits are done as shown by the curves in Fig. 1, where the short-dotted line and dashed line represent two separate Gaussian fits and solid lines represent the sum of the two Gaussian fits. From the fits to the relative momentum spectra as well as to the opening angular distributions, it is indicated that the component around q_{pp} =20 MeV/c occupies ~ 19.1% out of the whole q_{pp} spectra, and the one for θ_{pp} the component around 30° occupies 16.9% out of the whole θ_{pp} spectra, respectively. In the other words, the probability of diproton emission for ^{23}Al in this excitation energy window is about 18%.

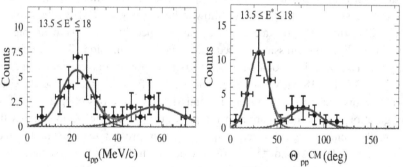

Fig. 2. Same as Fig. 1 but for the decay channel of p + p + ^{20}Ne from ^{22}Mg. The excitation energy window is 13.5≤E*≤18MeV. The dotted line and dashed line represent two separate Gaussian fits and solid line represents the sum of two Gaussian fits.

Similar analysis has been done for the even-even proton-rich nucleus ^{22}Mg. When the high excitation energy events are selected, the onset of diproton component occurs. Fig. 2 shows the relative momentum spectrum and opening angular distribution for the channel of ^{22}Mg → p + p + ^{20}Ne in the excitation energy window 13.5 ≤E*≤18 MeV. The peak of the relative momentum being at 20 MeV/c and the corresponding smaller opening angle is observed. This component is consistent with the diproton emission mechanism. Similar to the ^{23}Al case, two Gaussian fits are also used. From the fits, the component around q_{pp} =20 MeV/c occupies ~ 57.9% from the whole q_{pp} spectra and the one for θ_{pp} around 30° occupies 52.1% from the whole θ_{pp} spectra, respectively. In the other words, the probability of diproton emission for ^{22}Mg in this excitation energy window is about 54%.

From the data table of ^{23}Al, there exists an excited state of 11.780 MeV with J^{π}= 5/2^{+}, from which the two-proton can be emitted [21]. Even though our excitation energy data is not precise enough to identify the exact excited state, the selected excitation energy window 10.5 $\leq$E*$\leq$ 15MeV covers the 11.780 MeV excited state. Our observation illustrates two-proton emission is a possible decay channel which is consistent with above data sheet. However, since ^{23}Al is an odd-Z proton-rich nucleus, the pairing diproton emission is relatively difficult in comparison with the even-even proton-rich nucleus ^{22}Mg. Actually, our above data show that the probability of diproton emission of ^{23}Al is much less than that of ^{22}Mg. In previous beta-delayed proton emission experiment in the decay of ^{23}Al, two-proton emission has been established but the mechanism for that decay is uncertain [22]. In that experiment the transitions from the ^{22}Mg T=2 analog state (fed by the super-allowed beta decay of ^{23}Al) to the ground state and/or first excited state of ^{20}Ne was claimed but they were unable to distinguish diproton emission or sequential protons emission. The excitation energy of T=2 state is 14.044 MeV which is in region of our excitation energy window: 13.5$\leq$E*$\leq$18 MeV for ^{22}Mg. Our data confirm that there indeed exists diproton emission from single-step ^{2}He emission (two-protons coupled to a 1S$_{0}$ configuration) by the observation of the peak at q_{pp} = 20 MeV/c together with the small opening angles. Overall speaking, our present experiment for the proton-rich nuclei ^{23}Al and ^{22}Mg definitely demonstrates, for the first time, that there exists the component of diproton emission process from ^{2}He.

4. Summary

The measurements on two proton relative momentum together with the opening angle have been performed for the excited ^{23}Al and ^{22}Mg at RIKEN-RIPS beamline. In order to explore the internal proton-proton correlation information inside excited proton-rich nuclei, the decay channels of ^{23}Al $\rightarrow$ p + p + ^{21}Na and ^{22}Mg$\rightarrow$ p + p + ^{20}Ne have been analyzed. The excitation energy spectra are reconstructed by the invariant mass method for the above decay channels and the preliminary results on relative momentum and opening angle between two protons are presented. From our preliminary analysis, a peak around about 20 MeV/c for two-proton relative momentum distribution is clearly observed in higher excitation energy window together with small opening angle, which could be explained by the ^{2}He's diproton emission mechanism. Of course, sequential decay and/or three-body decay mechanisms for proton emissions are accompanied. For odd-Z pronton-rich nucleus ^{23}Al, the sequential decay is dominant. While, for even-even proton-rich nucleus ^{22}Mg, the diproton emission becomes favorable. More details on the analysis and the simulations are in progress.

Acknowledgment

The authors are very grateful to all of the staff at the RIKEN accelerator for providing beams during the experiment. The Chinese collaborators greatly appreciate the hospitality from the RIKEN-RIBS laboratory.

References

1. V. I. Goldansky, *Nucl. Phys.* **19,** 482 (1960); *Nucl. Phys.* **2,** 648 (1961); *Phys. Lett.* **14,** 233 (1965).
2. O. V. Bochkarev, L. V. Chulkov, A. A. Korsheninnikov, E. A. Kuzmin, I. G. Mukha, and G. B. Yankov, *Nucl. Phys.* A **505**, 215 (1989).
3. D. F. Geesaman, R. L. McGrath, P. M. S. Lesser, P. P. Urone, and B. VerWest, *Phys. Rev. C* **15** , 1835 (1977).
4. G. Raciti, G. Cardella, M. De Napoli, E. Rapisarda, F. Amorini, and C. Sfienti, *Phys. Rev. Lett.* **100,** 192503 (2008).
5. C. J. Lin, X. X. Xu, H. M. Jia et al., *Phys. Rev. C* **80,** 014310 (2009).
6. I. Mukha, K. S¨ummerer, L. Acosta et al., *Phys. Rev. Lett.* **99,** 182501 (2007).
7. B. Blank and M. Ploszajczak, *Rep. Prog. Phys.* **71,** 046301 (2008), and references therein.
8. G. Audi, A. H.Wapstra, and C. Thibault, *Nucl. Phys.* A **729,** 33 (2003).
9. X. Z. Cai et al., *Phys. Rev. C* **65,** 024610 (2002).
10. D. Q. Fang et al., *Phys. Rev. C* **76,** 031601(R) (2007).
11. A. Ozawa et al., *Phys. Rev. C* **74,** 021301(R) (2006).
12. V. E. Iacob et al., *Phys. Rev. C* **74,** 045810 (2006) .
13. A. Saastamoinen et al., *Phys. Rev. C* **80,** 044330 (2009).
14. T. Gomi, et al., *Nucl. Phys.* A **758,** 761c (2005).
15. M. Wiescher et al., *J. Phys. G* **25,** R133 (1999);
16. M. Wiescher, K. Langanke, *Z. Phys.* A **325,** 309 (1986).

158

17. S. Bishop *et al.*, *Phys. Rev. Lett.* **90**, 162501 (2003).
18. J. M. DAuria *et al.*, *Phys. Rev. C* **69**, 065803 (2004).
19. D. Seweryniak *et al.*, *Phys. Rev. Lett.* **94**, 032501 (2005).
20. B. Blank, *et al.*, *Phys. Rev. Lett.* **77**, 2893 (1996).
21. R. B. FIRESTONE, *Nuclear Data Sheets* **108**, 1 (2007).
22. M. D. Cable *et al.*, *Phys. Rev. Lett.* **50**, 404 (1983).

UNCERTAINTIES OF TH/EU AND TH/HF CHRONOMETERS FROM NUCLEAR MASSES

Z. M. NIU[1], B. SUN[2,3], and J. MENG[1,2,4*]

[1] *State Key Laboratory of Nuclear Physics and Technology, School of Physics, Peking University, Beijing 100871, China*
* E-mail: mengj@pku.edu.cn

[2] *School of Physics and Nuclear Energy Engineering, Beihang University, Beijing 100191, China*

[3] *Justus-Liebig-Universität Giessen, Heinrich-Buff-Ring 14, Giessen 35392, Germany*

[4] *Department of Physics, University of Stellenbosch, Stellenbosch, South Africa*

The age of the Universe is one of the most important physical quantities in cosmology and it can be determined with the r-process nucleochronometer. Based on the classical r-process model, the r-process abundance patterns and various nuclear chronometers in the metal-poor halo stars are investigated by employing the newly developed nuclear mass models. It is found that the uncertainty of Th/Eu chronometer caused by nuclear mass uncertainties is about 4 Gyr, while the uncertainty of Th/Hf chronometer is so large that it should be taken with caution. With the Th/Eu chronometer, the age of the metal-poor stars CS 31082-001 is determined as 16.3 ± 7.1 Gyr, which agrees well with the results derived from Th/U chronometer.

Keywords: Nuclear chronometer; r-process; metal-poor star.

1. Introduction

The age of the Universe is one of the most important physical quantities in cosmology. As the metal-poor stars were formed at the early epoch of the Universe, their ages can set a lower limit on the age of the Universe. The ages of these stars can be determined by nuclear chronometers, which rely on the comparison of the present abundances of radioactive nuclei with the initial abundances at their productions.

The initial abundance of these radioactive nuclei can be derived from r-process calculations. However, most of neutron-rich nuclei of relevance to the r-process are not accessible in experiments, consequently, r-process calculations crucially depend on accurate theoretical predictions for nuclear masses, β-decay half-lives, etc. Therefore, it is necessary to make a systematic investigation of the uncertainties of age estimate using nuclear masses. During the past decades, a number of investigations[1-4] in r-process chronometers have been reported in literatures based on a few widely used mass models, such as finite-range droplet model (FRDM).[5]

Recently, many mass models, such as Duflo-Zuker DZ31,[6] Hartree-Fock-Bogoliubov (HFB-17),[7] relativistic mean field (RMF),[8] and an improved macroscopic-microscopic mass formula with Woods-Saxon potential (WS*),[9] are developed. Although these newly developed mass models can well reproduce the experimental data, they are not included in the previous analysis of the nuclear chronometer. In Ref. 10, the influence of nuclear masses on Th/U chronometer is investigated with the newly developed mass models. However, the U lines were only detected in three metal-poor stars CS 31082-001,[11] BD +17°3248,[12] and HE 1523-0901.[13] Therefore, it is necessary to investigate the uncertainty of Th/X (X represents a stable element) chronometers, especially the commonly used Th/Eu and recently proposed Th/Hf[4] chronometers, using the nuclear mass predictions from these newly developed models.

In this work, we will analyze the abundance pattern of metal-poor star and deduce its age with the Th/Eu and Th/Hf chronometers. The improved nuclear mass models have been used in our r-process calculations. In Sec. 2, a brief introduction to the classical r-process model and the nuclear physics inputs used in this work are given. In Sec. 3, the influence of nuclear masses on the r-process calculations and the age estimate of the metal-poor star are discussed. Finally, a summary is presented in Sec. 4.

2. Sketch of the classical r-process model

The classical r-process model is adopted in this investigation to deduce the initial Th/Eu and Th/Hf abundance ratios as in Refs. 1,3,10. In this model, seed-nuclei (iron) are irradiated by neutron sources of high and continuous neutron densities n_n over a timescale τ in a high temperature environment ($T \sim 1$ GK).[14] A configuration of sixteen r-process components with neutron densities in the range of 10^{20} to 3×10^{27} cm^{-3} is chosen to reproduce the solar r-process abundances.[15-17] This model is considered as a realistic simplification of dynamical r-process model, and it has been successfully employed in describing r-process patterns of both the Solar System and metal-poor stars.[4,15,18,19]

In this work, available experimental data[20,21] are used, otherwise predictions of mass models DZ10,[22] DZ31,[6] FRDM,[5] HFB-17,[7] RMF,[8] and WS*[9] are employed. As for the β-decay rates, the predictions of the FRDM+quasiparticle random-phase approximation (QRPA) method[23] are employed throughout the paper as a complementary to the experimental data.[20]

3. Results and discussions

3.1. *The Influence of Mass Models*

The r-process calculations using various nuclear mass models can yield abundances differing even by several orders of magnitude.[15] It is therefore necessary to test their reliability before applying them to age estimates. In Fig. 1, the r-process simulations for DZ10, DZ31, FRDM, WS*, HFB-17, and RMF mass models are shown and compared with the observations. In the whole region of $38 \leqslant Z \leqslant 82$, the simulation using the DZ31 mass model better reproduce observations and the rms value with respect to observation is approximately 0.34, while the simulation using the RMF mass model has larger deviations for the heavier neutron-capture nuclei ($Z \geqslant 56$) and the rms value with respect to observations is 0.52. For the simulations using other mass models, the corresponding rms values are in between.

Since the elements Eu ($Z = 63$) and Hf ($Z = 72$) locate in the region $56 \leqslant Z \leqslant 82$, it is essential to well reproduce the observations in this region for the Th/Eu and Th/Hf chronometers. From Fig. 1, it is found that the simulations using the DZ10, WS*, DZ31, FRDM mass models better reproduce observations, and the rms values with respect to observations are 0.25, 0.31, 0.33, 0.38, respectively. Due to the large trough around $Z \sim 70$, the rms values increase to 0.48 and 0.58 for the simulations using HFB-17 and RMF mass models. This large trough might be due to the large neutron shell gap, which has not been well understood at present. Consequently, the abundances of Hf are underestimated for the simulations using HFB-17 and RMF mass models. Meanwhile, the underestimation of Hf abundance is also observed for the FRDM simulation, although the trough around $Z \sim 70$ is smaller for this simulation. This underestimation can lead to large scattering of Th/Hf abundance ratio, hence the age estimated from Th/Hf chronometer, when different mass models are employed for the r-process calculations.

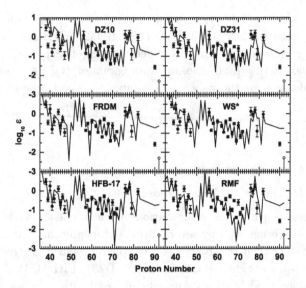

Fig. 1. Calculated r-process abundances (scaled to Eu) using various mass models. The filled circles represent the scaled average element abundances of CS 31082-001[24] and CS 22892-052[25] in the region $Z < 82$. The abundances for Pb, Th and the upper limit (open circle) on U are taken from CS 22892-052. We adopt the usual notation that $\log_{10} \varepsilon(A) \equiv \log_{10}(N_A/N_H) + 12.0$ for element A, where N represents abundance.

3.2. *Age Estimates from Th/Eu and Th/Hf Chronometers*

With the initial r-process abundance ratios Th/Eu and Th/Hf and their present observed values, one can eventually deduce the age of the metal-poor star. In Fig. 2, the ages of the metal-poor star CS 31082-001 are presented for different chronometers. For the Th/Eu chronometer, the left panel of Fig. 2, the ages determined using DZ10 and DZ31 mass models are similar, while the ages estimated using other mass models are slightly longer. Considering the errors from observations, all results using Th/Eu chronometer are agreement with each other. However, it is different for the Th/Hf chronometer. From the right panel of Fig. 2, it is found that the ages determined using FRDM, HFB-17, and RMF mass models are so large that they can not be consistent with the ages determined using DZ10 and DZ31 mass models. This deviation can be traced back to the large scattering of Th/Hf abundance (refer to Fig. 1). This might imply that the age estimate using Th/Hf chronometer should be taken with caution.

In order to estimate the age of a star more reliably, one could adopt the average values for different simulations. The corresponding uncertainty

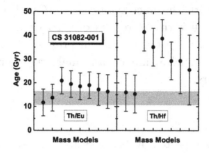

Fig. 2. Age of the metal-poor star CS 31082-001 determined using Th/Eu (left panel) and Th/Hf (right panel) chronometers. For each panel, the first six circles from the left to right represent the ages estimated using DZ10, DZ31, FRDM, HFB-17, RMF, and WS* mass models, respectively, while the last two circles denote the average ages with all mass models and the four mass models DZ10, DZ31, FRDM, and WS*. The shaded areas correspond to the age determined using the Th/U chronometer from Ref. 10.

is their root mean square deviation. From the above discussion, it is clear that the DZ10, DZ31, KTUY, and WS* simulations reproduce the stable element abundances better than the others. Therefore, these mass models might be more credible and are selected to estimate the age of a metal-poor star. For the Th/Eu chronometer, the age of metal-poor star CS 31082-001 is obtained as 16.3 ± 7.1 Gyr, which agrees with the value of 13.5 Gyr determined using Th/U chronometer in Ref. 10. The uncertainty of age determined here includes that of 4.3 Gyr from nuclear mass models and 5.6 Gyr from observation. Similarly, the age of CS 31082-001 from Th/Hf chronometer is deduced as 25.5 ± 14.7 Gyr. For comparison, the corresponding average ages for all the mass models using the Th/Eu and Th/Hf chronometers are determined as 17.2 ± 6.8 and 29.3 ± 13.8 Gyr with uncertainties of 3.6 and 11.3 Gyr from nuclear mass models, respectively. At last, it should point out that results calculated from Th/Hf chronometer should be taken with caution due to the large uncertainties of Th/Hf abundance ratio from nuclear masses.

4. Summary

In this work, the r-process abundance pattern in metal-poor star and the ages estimated using Th/Eu and Th/Hf chronometers have been investigated. It is found that the uncertainty of Th/Eu chronometer from nuclear masses is about 4 Gyr, while the uncertainty of Th/Hf chronometer is so large that it should be taken with caution. By adopting the Th/Eu chronometer, the age of the metal-poor stars CS 31082-001 is determined

164

as 16.3 ± 7.1 Gyr with the uncertainty of 4.3 Gyr from nuclear masses. This result agrees well with the results derived from Th/U chronometer. In order to reduce the age uncertainty in present work, it is essential to develop more reliable nuclear models and make more precise observations from metal-poor stars.

Acknowledgments

This work is partly supported by Major State 973 Program 2007CB815000 and the NSFC under Grant Nos. 10947013, 10947149 and 10975008, the NCET, and the Fundamental Research Funds for the Central Universities.

References

1. J. J. Cowan, B. Pfeiffer, K.-L. Kratz et al., Astrophys. J. **521**, 194 (1999).
2. S. Goriely, M. Arnould, Astron. Astrophys. **379**, 1113 (2001).
3. H. Schatz, R. Toenjes, B. Pfeiffer et al., Astrophys. J. **579**, 626 (2002).
4. K.-L. Kratz, K. Farouqi, B. Pfeiffer et al., Astrophys. J. **662**, 39 (2007).
5. P. Möller, J. R. Nix, W. D. Myers et al., Atom. Data Nucl. Data Tables **59**, 185 (1995).
6. A. P. Zuker, Rev. Mex. Fis. S **54**, 129 (2008).
7. S. Goriely, N. Chamel, J. M. Pearson, Phys. Rev. Lett. **102**, 152503 (2009).
8. L. S. Geng, H. Toki, J. Meng, Prog. Theor. Phys. **113**, 785 (2005).
9. N. Wang, M. Liu, X. Z. Wu, Phys. Rev. C **81**, 044322 (2010).
10. Z. M. Niu, B. H. Sun, J. Meng, Phys. Rev. C **80**, 065806 (2009).
11. R. Cayrel, V. Hill, T. C. Beers et al., Nature **409**, 691 (2001).
12. J. J. Cowan, C. Sneden, S. Burles et al., Astrophys. J. **572**, 861 (2002).
13. A. Frebel, N. Christlieb, J. E. Norris et al., Astrophys. J. **660**, L117 (2007).
14. K.-L. Kratz, J. P. Bitouzet, F.-K. Thielemann et al., Astrophys. J. **403**, 216 (1993).
15. B. Sun, F. Montes, L. S. Geng et al., Phys. Rev. C **78**, 025806 (2008).
16. B. Pfeiffer, K.-L. Kratz, F.-K. Thielemann, Z. Phys. A **357**, 235 (1997).
17. B. Sun, J. Meng, Chin. Phys. Lett. **25**, 2429 (2008).
18. C. Freiburghaus, J.-F. Rembges, T. Rauscher et al., Astrophys. J. **516**, 381 (1999).
19. B. Pfeiffer, K.-L. Kratz, F.-K. Thielemann et al., Nucl. Phys. A **693**, 282 (2001).
20. G. Audi, O. Bersillon, J. Blachot et al., Nuclear Physics A **729**, 3 (2003).
21. B. Sun, R. Knöbel, Yu. A. Litvinov et al., Nuclear Physics A **812**, 1 (2008).
22. J. Duflo, A. P. Zuker, http://theory.gsi.de/~ksieja/ssp/DZ/du_zu_10.feb96.
23. P. Möller, B. Pfeiffer, K.-L. Kratz, Phys. Rev. C **67**, 055802 (2003).
24. V. Hill, B. Plez, R. Cayrel et al., Astron. Astrophys. **387**, 560 (2002).
25. C. Sneden, J. J. Cowan, J. E. Lawler et al., Astrophys. J. **591**, 936 (2003).

THE CHIRAL DOUBLET BANDS WITH $\pi(g_{9/2})^{-1} \otimes \nu(h_{11/2})^2$ CONFIGURATION IN A∼100 MASS REGION

B. QI*, S. Y. WANG, C. LIU, C. J. XU, L. LIU and D. P. SUN

*School of Space Science and Physics, Shandong University at Weihai,
Weihai, 264209, China*
E-mail: bqi@sdu.edu.cn

The chiral doublet bands with $\pi(g_{9/2})^{-1} \otimes \nu(h_{11/2})^2$ configuration are studied by the triaxial particle rotor model. The energy spectra and the in-band $B(M1)/B(E2)$ of the doublet bands with different triaxiality parameter γ are analyzed.

Keywords: Chiral doublet bands; $A \sim 100$ mass region; triaxiality parameter.

1. Introduction

Since the prediction of chirality in atomic nuclei in 1997,[1] much effort has been devoted to explore this interesting phenomenon. So far, more than 20 candidate chiral nuclei have been reported experimentally in the $A \sim 100$, 130 and 190 mass regions. An overview of these studies and open problems in understanding the nuclear chirality is introduced in Ref. 2.

In $A \sim 100$ mass region, not only in odd-odd nuclei, but also in odd-A [103]Rh[3] and [105]Rh,[4] the existence of chiral doublet bands has been reported. The corresponding quasiparticle configurations are suggested as $\pi(g_{9/2})^{-1} \otimes \nu(h_{11/2})^1$ for odd-odd nuclei [5,6] and $\pi(g_{9/2})^{-1} \otimes \nu(h_{11/2})^2$ for odd-A nuclei.[3,4] On the theoretical side, various approaches have been applied to study the nuclear chirality in odd-odd nuclei in $A \sim 100$ mass region.[7,8] However, the theoretical study for odd-A nuclei in $A \sim 100$ mass region was seldom. Such study for the doublet bands can be carried out by either triaxial particle rotor model (PRM) with n-particle-n-hole configurations[9] or random phase approximation (RPA) calculations based on the titled cranking model (TAC).[10] In Ref. 9, we have developed the triaxial n-particle-n-hole PRM and applied the model to study the nuclear chirality in odd-A nucleus [135]Nd. It is therefore interesting to further study the doublet bands in odd-A nuclei with $A \sim 100$ in the framework of the one

$g_{9/2}$ proton hole and two $h_{11/2}$ neutron particles coupled to triaxial rotor model.

Thus, the chiral doublet bands with $\pi(g_{9/2})^{-1} \otimes \nu(h_{11/2})^2$ configuration will be studied via a triaxial 2-particle-1-hole PRM. In particular, the energy spectra and $B(M1)/B(E2)$ ratios of the doublet bands for different triaxiality parameter γ will be discussed here.

2. Discussions

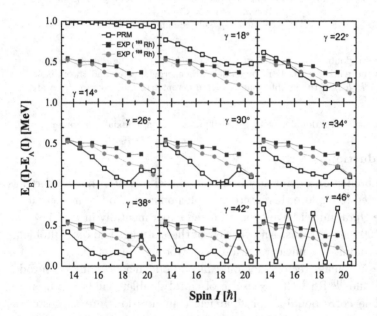

Fig. 1.   Energy difference between the two lowest bands A and B calculated in PRM with different triaxiality parameters γ with configuration $\pi g_{9/2}^{-1} \otimes \nu h_{11/2}^2$ (open symbols) in comparison with the data of ^{103}Rh (filled squares) and ^{105}Rh (filled dots).

The detailed formulae for the particle rotor model which couples 1 proton hole and 2 neutron particles to a rigid triaxial rotor are described in Ref. 9. In the present calculations, the configuration $\pi(g_{9/2})^{-1} \otimes \nu(h_{11/2})^2$ is adopted. The quadrupole deformation parameter $\beta = 0.23$ is obtained from the microscopic self-consistent triaxial relativistic mean field calculation for this mass region.[11] The moment of inertia takes value of 21 MeV$^{-1}\hbar^2$ according to the experimental energy spectra in this mass region. For the electromagnetic transition, the empirical intrinsic quadrupole moment and gyromagnetic ratios[12] $g_R = 0.44$, and $g_p = 1.26$, $g_n = -0.21$ are adopted.

The energy differences between the two lowest bands (denoted by A and B) calculated in PRM with different triaxiality parameter γ are shown in Fig. 1. As the γ value increases from $14°$ to $30°$, the energy difference between doublet bands decreases rapidly, namely the good degeneracy for band A and B is achieving. When γ value takes value of $42°$ or $46°$, the staggering of energy difference is very obvious, namely the signature splitting is very obvious for $42°$, $46°$ case. For most values, the smallest energy differences are obtained at spin $37/2\hbar$. At this point, the energy differences are 946, 475, 175, 21, 34, 93, 126, 60, 43 keV for $\gamma = 14°, 18°, 22°, 26°, 30°, 34°, 38°, 42°, 46°$ respectively. In chiral picture, the observation of two almost degenerate $\Delta I=1$ rotational bands is considered as the fingerprint of chiral doublet bands.[1] Thus the $14°$ case with large energy difference between the doublet bands is not consistent with the chiral hypothesis. Furthermore, the energy spectra should possess a smooth dependence with spin in chiral picture.[5] Thus the $42°, 46°$ cases with obvious signature splitting should also not be considered as the chiral doublet bands. In a word, from the properties of energy spectra the $18° \leq \gamma \leq 38°$ cases in Fig. 1 could be considered as chiral doublet bands.

The in-band $B(M1)/B(E2)$ values of the two lowest bands A and B with $\pi(g_{9/2})^{-1} \otimes \nu(h_{11/2})^2$ configuration are also calculated in PRM. These calculated in-band $B(M1)/B(E2)$ values are sensitive to the triaxiality parameter γ. The calculated in-band $B(M1)/B(E2)$ ratios in the doublet bands are almost the same over the whole spin region for $18° \leq \gamma \leq 38°$ cases, while are very different for $42°, 46°$ cases. In ideal chiral doublet bands, it is pointed out that all corresponding properties must be identical or, in practice, very similar. Thus the very similar $B(M1)/B(E2)$ in the doublet bands is a fingerprint for ideal chiral doublet bands. From this consideration, the $42°, 46°$ case are also excluded from the chiral hypothesis.

3. Conclusions

In summary, the doublet bands with $\pi(g_{9/2})^{-1} \otimes \nu(h_{11/2})^2$ configuration are discussed in the particle rotor model for different triaxiality parameter γ. Based on these calculations, it is found that γ is a sensitive parameter for the properties of these doublet bands. It is found that the $18° \leq \gamma \leq 38°$ cases could be considered as the chiral doublet bands, while $14°, 42°, 46°$ cases in our calculations should be excluded from the chiral hypothesis.

Acknowledgements

This work is supported by the National Natural Science Foundation (Grant Nos. 11005069 and 10875074), and the Shandong Natural Science Foundation (Grant No. ZR2010AQ005).

References

1. S. Frauendorf and J. Meng, *Nucl. Phys. A* **617**, 131 (1997).
2. J. Meng and S. Q. Zhang, *J. Phys. G* **37**, 064025 (2010).
3. J. Timár *et al.*, *Phys. Rev. C* **73**, 011301(R) (2006).
4. J. Timár *et al.*, *Phys. Lett. B* **598**, 178 (2004).
5. C. Vaman, D. B. Fossan, T. Koike, K. Starosta, I. Y. Lee, and A. O. Macchiavelli, *Phys. Rev. Lett.* **92**, 032501 (2004).
6. P. Joshi *et al.*, *Phys. Lett. B* **595**, 135 (2004).
7. S. Y. Wang, S. Q. Zhang, B. Qi, J. Peng, J. M. Yao, and J. Meng, *Phys. Rev. C* **77**, 034314 (2008).
8. J. Peng, J. Meng and S. Q. Zhang, *Chin. Phys. Lett.*, **20**, 1123 (2003).
9. B. Qi, S. Q. Zhang, J. Meng, and S. Frauendorf, *Phys. Lett. B*, **675**, 175 (2009).
10. S. Mukhopadhyay *et al.*, *Phys. Rev. Lett.* **99**, 172501 (2007).
11. J. Meng, J. Peng, S. Q. Zhang, and S.-G. Zhou, *Phys. Rev. C* **73**, 037303 (2006).
12. B. Qi, S. Q. Zhang, S. Y. Wang, J.M. Yao and J. Meng, *Phys. Rev. C* **79**, 041302(R) (2009).

α FORMATION PROBABILITIES IN NUCLEI AND PAIRING COLLECTIVITY

CHONG QI

KTH, Alba Nova University Center, SE-10691 Stockholm, Sweden
E-mail: chongq@kth.se

α formation amplitudes extracted from experimental data are presented and an abrupt change around the $N = 126$ shell closure is noted. It is explained as a sudden hindrance of the clustering of nucleons. The clustering induced by the pairing mode acting upon the four nucleons is inhibited if the configuration space does not allow a proper manifestation of the pairing collectivity.

Keywords: Alpha decay; formation amplitudes; pairing collectivity.

It is nearly a century ago that the Geiger-Nuttall law was formulated based on α decay systematics,[1] which was reproduced nicely by the Gamow theory.[2] A proper calculation of the decay process needs to address firstly the clustering of the nucleons at a distance outside the nuclear surface and secondly the evaluation of the penetrability of the cluster through the Coulomb and centrifugal barriers. The description of the cluster in terms of its components requires a microscopic many-body framework. The importance of a proper treatment of α decay was attested by a recent calculation.[3,4] One thus obtained a generalization of the Geiger-Nuttall law which holds for all isotopic chains and all cluster radioactivities. In this universal decay law (UDL) the penetrability is still a dominant quantity. By using three parameters only, one finds that all known ground-state to ground-state radioactive decays are explained rather well. This good agreement is a consequence of the smooth transition in the nuclear structure that is often found when going from a nucleus to its neighboring nuclei. This is also the reason why, e.g., the BCS approximation works so well in many nuclear regions.

In this work we show that, when a sudden transition occurs in a given chain of nuclei, departures from the UDL can be seen.[5] Our aim is to understand why this difference appears. We will also try to discern whether one can, in general, obtain information about the structure of the nuclei involved in the decay.

The classical expression for the decay width Γ_l is written as,[6]

$$\Gamma_l(R) = 2\mathcal{P}_l(R)\frac{\hbar^2}{2\mu R}|\mathcal{F}_l(R)|^2, \qquad (1)$$

where l is the angular momentum carried by the outgoing α-particle, $\mathcal{P}$ is the penetration probability and μ is the reduced mass of the final system. R is the radius where the wave function of the α-particle formed in the internal region is matched with the outgoing two-body wave function in the external region. The amplitude of the wave function in the internal region is the formation amplitude, i.e.,

$$\mathcal{F}_l(R) = \int d\mathbf{R}d\xi_d d\xi_\alpha [\Psi(\xi_d)\phi(\xi_\alpha)Y_l(\mathbf{R})]^*_{J_m M_m}\Psi_m(\xi_d,\xi_\alpha,\mathbf{R}), \qquad (2)$$

where d, α and m label the daughter, α particle and mother nuclei, respectively. Ψ are the intrinsic wave functions and ξ the corresponding intrinsic coordinates. $\phi(\xi_\alpha)$ is a Gaussian function of the relative coordinates of the two neutrons and two protons that constitute the α-particle.[7,8] The rest of the notation is standard.

The main problem in the evaluation of formation amplitude is the description of the clusterization of the four nucleons that eventually become the α-particle. The first calculations were performed after the appearance of the shell model. These calculations had limited success due to the small shell model spaces that could be included at that time.[7] The role of configuration mixing in inducing clustering was shown much later[8,9] in the case of the decay of the nucleus ^{212}Po. In pursuing this task one has found that the mode that determines clusterization is the pairing vibration.[8,9]

The experimental decay half-lives[10] can be compared with the ones predicted by the UDL[3] by using the ratio between the two. For clarity of presentation we will define the ratio R ($R \geq 1$) as $\log_{10} R = |\log_{10} T^{\text{Expt.}}_{1/2} - \log_{10} T^{\text{UDL}}_{1/2}|$. In all decay cases analyzed in this paper the α particle carries angular momentum $l = 0$. From Fig. 1 one notes that in the range $82 \leq Z \leq 86$, that is between Pb and Rn isotopes, the UDL underestimate the experimental α decay half-lives by large factors. In all cases plotted in the figure where large differences between experimental data and the UDL are seen (i.e., with $R \geq 3$), the decays are hindered with respect to the average values predicted by the UDL.

The abrupt changes in the α decay cases analyzed above can be discerned even more clearly by looking at the absolute values of the formation

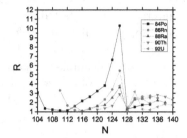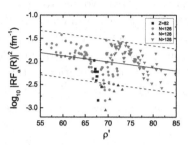

Fig. 1. Left: Discrepancy between experimental decay half-lives and UDL calculations as a function of the neutron number of the mother nucleus N; Right: $\log_{10} |RF(R)|^2$ as a function of ρ'.[5]

amplitudes. Using the experimental decay half-lives[10] one can extract the formations amplitudes as

$$\log_{10} |R\mathcal{F}_\alpha(R)| = -\frac{1}{2} \log_{10} T_{1/2}^{\text{Expt.}} + \frac{1}{2} \log_{10} \left[\frac{\ln 2}{\nu} |H_0^+(\chi, \rho)|^2 \right]. \quad (3)$$

This is shown in Fig. 1 for different even-even isotopes as a function of the quantity $\rho' = \sqrt{2\mathcal{A}Z_d(A_d^{1/3} + 4^{1/3})}$, where $\mathcal{A} = 4A_d/A_m$.[3]

One notices in Fig. 1 that for most cases the UDL predicts the experimental values within a factor of three, except for cases around $N = 126$. The case that shows the most significant hindrance corresponds to the α decay of the nucleus ^{210}Po. The symbols with $\log_{10} |R\mathcal{F}_\alpha(R)|^2 \sim -2.7$ fm^{-1} correspond to the α decays of nuclei ^{208}Po, ^{212}Rn and ^{194}Pb. Finally, it is worthwhile to point out that this sudden change in α-decay systematics at $N = 126$ has also been noticed in Refs.[11-14]

We will analyze the α formation amplitudes of the isotopes that show the kink at $N = 126$ discussed above. We will take the decay of ^{210}Po as a typical example and compare it with that of ^{212}Po. To compare with experimental data, we extract the magnitude of the formation amplitudes from measured half-lives by using Eq. (3). One thus obtains the value $\mathcal{F}_\alpha(R) = 3.305 \times 10^{-3}$ fm$^{-3/2}$ in ^{210}Po and $\mathcal{F}_\alpha(R) = 1.082 \times 10^{-2}$ fm$^{-3/2}$ in ^{212}Po. These correspond to a variation in the formation amplitudes by a factor of 3.28, that is a factor of 10.73 in the formation probabilities.

Within the shell model a four-particle state α_4 in ^{212}Po can be written as

$$|^{212}\text{Po}(\alpha_4)\rangle = \sum_{\alpha_2 \beta_2} X(\alpha_2 \beta_2; \alpha_4) |^{210}\text{Pb}(\alpha_2) \otimes^{210} \text{Po}(\beta_2)\rangle \quad (4)$$

where α_2 (β_2) labels two-neutron (two-proton) states. The amplitudes X are influenced by the np interaction. If this interaction is neglected, then only one of the configurations in Eq. (4) would appear. This is done, for instance, in cases where the correlated four-particle state is assumed to be provided by collective vibrational states. Rather typical examples of such states are $|^{210}\text{Pb(gs)}\rangle$ and $|^{210}\text{Po(gs)}\rangle$. It is therefore calculations have been performed by assuming that $|^{212}\text{Po(gs)}\rangle$ is a double pairing vibration,[8,9] i.e.,

$$|^{212}\text{Po(gs)}\rangle = |^{210}\text{Pb(gs)} \otimes {}^{210}\text{Po(gs)}\rangle. \qquad (5)$$

The corresponding formation amplitude acquires the form,

$$\mathcal{F}_\alpha(R;{}^{212}\text{Po}) = \int d\mathbf{R}d\xi_\alpha \phi_\alpha(\xi_\alpha)\Psi(\mathbf{r_1 r_2};{}^{210}\text{Pb(gs)})\Psi(\mathbf{r_3 r_4};{}^{210}\text{Po(gs)}), \qquad (6)$$

where $\mathbf{r_1}, \mathbf{r_2}$ $(\mathbf{r_3}, \mathbf{r_4})$ are the neutron (proton) coordinates and $\mathbf{R}$ is the center of mass of the α particle.

With this expression for the formation amplitude the experimental half-life is reproduced rather well if a large number of high-lying configurations is included. Our calculations are done by using a surface delta interaction and nine major shells of a harmonic oscillator (HO) representation.[5]

The decay of the nucleus $^{210}\text{Po(gs)}$ leads to the daughter nucleus $^{206}\text{Pb(gs)}$, which is a two-hole state. The formation amplitude becomes,

$$\mathcal{F}_\alpha(R;{}^{210}\text{Po}) = \int d\mathbf{R}d\xi_\alpha \phi_\alpha(\xi_\alpha)\Psi(\mathbf{r_1 r_2};{}^{206}\text{Pb(gs)})\Psi(\mathbf{r_3 r_4};{}^{210}\text{Po(gs)}). \qquad (7)$$

By comparing Eqs. (6) and (7) one sees that the only difference between the two expressions is the two-neutron wave function, which corresponds to the two-particle state $^{210}\text{Pb(gs)}$ in Eq. (6) and to the two-hole state $^{206}\text{Pb(gs)}$ in Eq. (7). Therefore the kink observed experimentally should be related to the difference in clusterization induced by the pairing force in these two cases. To analyze the clustering features we will consider only the spin-singlet component, i.e., $(\chi_1\chi_2)_0$, of the two-body wave function, since that is the only part entering the intrinsic α-particle wave function. This component has the form,

$$\Psi_2(r_1, r_2; \theta_{12}) = \frac{1}{4\pi}\sum_{p\leq q}\sqrt{\frac{2j_p + 1}{2}}X(pq; \text{gs})\varphi_p(r_1)\varphi_q(r_2)P_{l_p}(\cos\theta_{12}), \qquad (8)$$

where φ is the single-particle wave function and P_l is the Legendre polynomial of order l satisfying $P_l(\cos 0) = 1$ (notice that for the ground states studied here it is $l_p = l_q$). As mentioned above, the pairing vibrations show strong clustering features as the number of single-particle states is increased.[9] But another manifestation of the pairing collectivity is an enhancement of the wave function on the nuclear surface. The reason of this enhancement is that all configurations contribute with the same phase in the building up of the two-particle wave function on the nuclear surface. The same mechanism increases the α formation amplitude and, therefore, the relative values of the wave functions of ^{210}Pb(gs), ^{210}Po(gs) and ^{206}Pb(gs) on the nuclear surface give a measure of the importance of the corresponding formation amplitudes.

To study the behavior of the two-particle wave functions we will apply Eq. (8) with $r_1 = 9$ m. We have plotted in Fig. 2 $\Psi_2(r_1, r_2, \theta)$ as a function of r_2 and θ. One sees that the wave functions are indeed strongly enhanced at the nuclear surface, as expected. But the important feature for us is that the enhancement is strongest in ^{210}Pb(gs) and weakest in ^{206}Pb(gs). This is because there is a relatively small number of configurations in the hole-hole case. In addition, the radial wave functions corresponding to the high-lying particle states extend farther out in space with respect to the hole configurations.

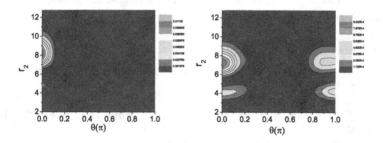

Fig. 2. The square of the two-boday wave function $|\Psi_2(r_1, r_2, \theta)|^2$ with $r_1 = 9$ fm for the two neutrons in ^{210}Pb (left) and two neutron holes in ^{206}Pb (right).

With these two-body wave functions we proceeded to evaluate the α formation amplitudes in ^{212}Po(gs) and ^{210}Po(gs).[5] One thus finds that with $R = 9$ fm the observed ratio between the formation amplitudes in ^{212}Po and ^{210}Po can be reproduced nicely.

In summary, the recently proposed universal decay law (UDL)[3] has been applied to perform a systematic calculation of α decay half-lives over all experimentally known cases. We found that although the UDL reproduces nicely most available experimental data, as expected, there is a case where it fails by a large factor. This corresponds to the α decays of nuclei with neutron numbers equal to or just below $N = 126$.[5] The reason for this large discrepancy is that in $N \leq 126$ nuclei the α formation amplitudes are much smaller than the average quantity predicted by the UDL (Fig. 1).

The case that shows the most significant hindrance corresponds to the α decay of the nucleus ^{210}Po. Starting from the formal definition of Eq. (2), we calculated the α formation amplitude of ^{210}Po and compared it with that of ^{212}Po. In these two cases the formation amplitudes can be described by the simple expressions (6) and (7). We found that the formation amplitude in ^{210}Po is hindered with respect to the one in ^{212}Po due to the hole character of the neutron states in the first case. This is a manifestation of the mechanism that induces clusterization, which is favored by the presence of high-lying configurations. Such configurations are more accessible in the neutron-particle case of ^{212}Po than in the neutron-hole case of ^{210}Po. This is a general feature in nuclei where neutrons and protons occupy different low-lying major shells.

Acknowledgment

This work has been supported by the Swedish Research Council under grant Nos. 623-2009-7340 and 2010-4723.

References

1. H. Geiger and J. M. Nuttall, *Philos. Mag.* **22**, 613 (1911); H. Geiger, *Z. Phys.* **8**, 45 (1922).
2. G. Gamow, *Z. Phys.* **51**, 204 (1928).
3. C. Qi, F. R. Xu, R. J. Liotta, and R. Wyss, *Phys. Rev. Lett.* **103**, 072501 (2009).
4. C. Qi *et al.*, *Phys. Rev. C* **80**, 044326 (2009).
5. C. Qi *et al.*, *Phys. Rev. C* **81**, 064319 (2010).
6. R. G. Thomas, *Prog. Theor. Phys.* **12**, 253 (1954).
7. H. J. Mang, *Phys. Rev.* **119**, 1069 (1960).
8. I. Tonozuka and A. Arima, *Nucl. Phys.* **A323**, 45 (1979).
9. F. A. Janouch and R. J. Liotta, *Phys. Rev. C* **27**, 896 (1983).
10. G. Audi, O. Bersillon, J. Blachol, and A. H. Wapstra, *Nucl. Phys.* **A729**, 3 (2003).
11. K. S. Toth *et al.*, *Phys. Rev. C* **60**, 011302(R) (1999).
12. J. Wauters *et al.*, *Phys. Rev. C* **47**, 1447 (1993).
13. K. Van de Vel *et al.*, *Phys. Rev. C* **68**, 054311 (2003).
14. D. N. Poenaru, R. A. Gherghescu, and N. Carjan, *Eur. Phys. Lett.* **77**, 62001 (2007).

A THEORETICAL PERSPECTIVE ON TRIGGERED GAMMA EMISSION FROM ^{178}Hf^{m2} ISOMER

SHUIFA SHEN[1,2,3,4]* and YUPENG YAN[3,4]

[1]*Key Laboratory of Radioactive Geology and Exploration Technology Fundamental Science for National Defense, East China Institute of Technology, Fuzhou, Jiangxi 344000, China*
*E-mail: shfshen@ecit.cn
www.ecit.edu.cn

[2]*International Centre for Materials Physics, Chinese Academy of Sciences, Shenyang 110016, China*

[3]*School of Physics, Suranaree University of Technology, Nakhon Ratchasima 30000, Thailand*

[4]*ThEP Center, Commission on Higher Education, 328 Si Ayutthaya Road, Ratchathewi, Bangkok 10400, Thailand*

TIANLI YANG[5]** and FANHUA HAO[5]

[5]*Institute of Nuclear Physics and Chemistry of CAEP, Mianyang, Sichuan 621900, China*
**E-mail: yangtianlilj@hotmail.com

HONGLIANG LIU, JIANYU ZHU, CHANGFENG JIAO and FURONG XU

School of Physics and State Key Laboratory of Nuclear Physics and Technology, Peking University, Beijing 100871, China

A number of applications have been proposed concerning the isomers, including the creation of a gamma-ray laser, since some of them may store large amounts of energy for long times. Against this background, the field of triggered gamma emission is entering a new phase in which improved level data allow targeting of specific potentially-useful transitions. In the present work, the configuration-constrained (diabatic blocking) calculations, with inclusion of γ-deformation, are performed to study the multi quasi-particle (multi-qp) excitations of ^{178}Hf. The detailed excitation energies known in ^{178}Hf are reasonably reproduced. From our calculations, two levels besides the existing level at 2.573 MeV are found for candidates of these intermediate states (or called depletion levels). Whether these two excited states actually exist will require additional experimentation.

Keywords: Triggered gamma emission; nuclear isomers; multi-quasiparticle states; potential energy surfaces.

1. Introduction

Long-lived high-energy nuclear isomeric states are suitable media for storing the energy. The controlled release of energy stored in nuclear isomeric samples will produce powerful pulsed-sources of gamma ray radiation. That is why, at present, this field is at the focus of a great scientific and technological interest. In addition, the possibility that the gamma bursts produced to have coherence and directionality (gamma-ray lasers) brings even more attention. It would be the very best of all possibilities to be able to immediately realize the release of high energy densities into a coherent radiation field. However, simply the induced emission of gamma rays from an isomeric sample would be of considerable technological significance. The triggering of the gamma radiation has been the object of our research as a first step in the pumping of a gamma ray laser. Many efforts and concepts have been put forward to find the best isomeric candidates and solutions to control the release of their energy. The top position of the most interesting candidates list is occupied by $^{178}\mathrm{Hf}^{m2}$ isomer[1,2] because it is long lived, available in research quantities, has a well known decay scheme, high excitation energy, a stable ground state, and targets of the isomeric Hf can be fabricated. The isomeric level lies at high excitation energy where high density of states exists; some of these states may have accentuated K-mixing character, which is the most important requirement of a trigger level. But up to now the induced γ emission in the case of spin-16, 2.447 MeV isomer in $^{178}\mathrm{Hf}$ is not established. The works carried out by Collins and coworkers[3–5] presented positive results, but they were not reproduced by much more elaborated experiments.[6–9] However, it is interesting to study the process in the case of 16^+ isomer of $^{178}\mathrm{Hf}$ from the nuclear physics point of view. By the way, although their experimental results and theoretical estimates did not support the low energy depletion of $^{178}\mathrm{Hf}^{m2}$ isomer, Carroll et al.,[9] did not deny the existence of lower-lying intermediate states (relative to the isomer).

The two-step process has been simply assumed in the triggered gamma-ray emission previously (see Fig. 1 of Ref. [10]). A nucleus in an isomeric state is first excited to an intermediate state (or called depletion level) by absorption of an incident photon. An intermediate state may serve as a "gateway" that connects the isomer to the ground state (g.s.). Level schemes like that reported in Ref. [11] provide important guidance to known electromagnetic transitions that might enable triggering. It was determined that the 14^-, 68-μs isomer (with $K = 14$) decayed with several branches, one of which led to the 8^-, 4-s isomer band and another of which led to the 31-year, 16^+ isomer ($K=16$). The essential point is that the latter transition has an $M2$ character and with $L=2$ and $\Delta K=2$, there is no K hindrance. Thus, the 126 keV transition from the 16^+ isomer to the 14^- isomer should produce triggered gamma emission in a cascade that closely resembles the natural decay of the 31-year state. The efficiency of depletion of $^{178}Hf^{m2}$ using incident 126 keV photons would be very low, so it is perhaps of less value for potential applications, but this would be a valuable scientific step.

In determining the favorable conditions for triggering gamma-ray emission from the isomer, it is necessary to have information on the structure of possible gateway states, as well as possible paths of electromagnetic transitions to and from these states. Experimentally, only a few states close to the ^{178}Hf isomer have been observed and documented in the literature.[12,13] The purpose of the present work is to demonstrate that the potential energy surfaces (PES) calculations[14] is an appropriate theory for studying high spin isomers and associated excitations, including potential gateway states; and to show that in the ^{178}Hf case, in order to trigger emission from the 16^+ isomer with low-energy photons, there are a few gateway states that may be supplied from our theoretical consideration.

2. Multi-quasiparticle potential-energy surfaces calculations

The configuration-constrained (diabatic-blocking) PES calculations are performed within the non-axial deformed Woods-Saxon (WS) potential[15] with universal parameters.[16] The monopole pairing strength parameter G is firstly determined by the average-gap method,[17] then adjusted to include the shape and blocking effects on odd-even mass differences.[14] In order to avoid the spurious phase transition encountered in the BCS approach, we employ the approximate particle-number projection by means of the Lipkin-Nogami (LN) method.[18,19] As an improvement, present calculations include quadrupole pairings which were not taken into account in the earlier works (see, for example, Ref. [14]). The strengths of the quadrupole pairing are determined by restoring the local Galilean invariance[20] of the system. Quadrupole pairing in doubly stretched coordinate space[20] has a negligible effect on energies, but it is included because it has an important influence on collective angular momenta.[21] Thus both collective and noncollective modes can be treated in a consistent theoretical framework.[22] Hence, no parameter is re-adjusted. The total energy of a configuration consists of a macroscopic part which is obtained by the standard liquid-drop model with the original parameters of Ref. [23] and a microscopic part resulting from the Strutinsky shell correction approach,[18,24] $\delta E_{shell} = E_{LN} - \tilde{E}_{Strut}$, including blocking effects. The excitation energy of a multi-quasiparticle (multi-qp) state is the total-energy difference between the excited and ground states. The constrained configuration energy in the LN approach, which contains the extra Lagrange multiplier, λ_2, is given by

$$E_{LN} = \sum_{j=1}^{s} e_{k_j} + \sum_{k \neq k_j} 2V_k^2 e_k - \Delta^2/G - G \sum_{k \neq k_j} V_k^4$$
$$+ G(N - S)/2 - 4\lambda_2 \sum_{k \neq k_j} (U_k V_k)^2, \qquad (1)$$

with

$$N - S = \sum_{k \neq k_j} 2V_k^2, \qquad (2)$$

where S is the proton or neutron seniority for a given configuration, i.e., the number of the blocked orbitals (indicated by index k_j), and N is the proton or neutron number. The blocking effect is achieved by removing the singly occupied orbitals from the LN calculation. The PES's are calculated in the lattice of quadrupole (β_2, γ) deformations with hexadecapole (β_4) variation. Since an intrinsic (non-rotation) PES is reflection-symmetric about $\gamma=0°$ and -60°, the absolute value $|\gamma|$ is used in this work. Note also that the configuration-constrained PES calculation requires a limitation to the range of the γ variation, because the WS basis will change when calculations go across the triaxial γ lines. Usually, the range of -30°$<\gamma<$30° is taken for prolate cases,[25] and -90°$<\gamma<$-30° for oblate cases.

A realistic treatment of a multi-qp state can be achieved with a configuration-constrained calculation of the potential-energy surface,[14] i.e. when changing deformation, to always follow and block the given orbitals which are occupied by the specified quasiparticles (this is the meaning of "diabatic blocking"). It appears that the calculated levels at 15^+, 2.677 and 14^-, 2.837 MeV lie between two levels at 2.405 (corresponds to 31-year, $K^\pi=16^+$ isomer at 2.447 MeV) and 3.096 MeV (corresponds to $K^\pi=14^-$ isomer at 2.573 MeV), so these two levels besides the $K^\pi=14^-$ isomeric state are the most likely candidates for photodeexcitation of the spin-16, 31-year isomer. But many important parameters are as yet unknown, and they can be evaluated only after the new experiments carried out.

The calculations carried out by Sun et al.,[10] suggest that ^{178}Hf exhibits significant γ softness already near its ground state. But in the present work the calculated PES's for the g.s., observed $K^\pi =16^+$ isomeric state and all other multi-qp states suggest a rigidity to axially asymmetric shapes (see, for example, Fig. 1), which is consistent with the result deduced by

Andrejtscheff and Petkov using the sum-rule method applying approximations[26] and with the result obtained by Hinke et al.[27] (see Fig. 8 of Ref.
[27]). The rigidity of the ground state is evidenced by the existence of the
relatively high-lying γ band, whose band head lies at 1175 keV.[13] Furthermore in general the multi-qp states are more rigid than the ground state.

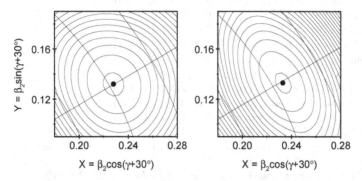

Fig. 1. Calculated PES's for the g.s. (in left) and $K^\pi = 16^+$ isomer (2-q$\nu \otimes$2-qπ in
right) for ^{178}Hf. The black dot indicates the lowest minimum, and the energy difference
between neighboring contours is 200 keV.

Triggered emission of gamma rays, releasing the energy stored in long-
lived nuclear metastable states, has been experimentally established since
1987 for the depletion of the 10^{15}-year isomer of ^{180}Ta.[28] In the near future,
we hope that the triggered gamma emission may be conclusively demonstrated for ^{178}Hf studied in the present work and for other long-lived isomers, such as those of ^{177}Lu and ^{242}Am. Whether they can occur, only
experiments will tell.

3. Summary

In summary, we have performed deformation-pairing self-consistent calculations for multi-qp states, with the following results: (i) All multi-qp states
and the ground state in ^{178}Hf exhibit an axially symmetric deformed shape
with almost the same quadrupole deformation β_2 and are rigid to axially asymmetric shapes. (ii) Two levels besides the already known level at
2.573 MeV are found for candidates of the potential gateway states. However, transition-rate calculations are beyond the scope of the present work
and are assumed to be very low.[9] Thus, improved experimental methods
by which to better examine these claims will be valuable.

References

1. C.B. Collins and J.J. Carroll, *Hyperfine Interact.* **107**, 3 (1997).
2. J. J. Carroll *et al.*, *Hyperfine Interact.* **135**, 3 (2001).
3. C. B. Collins *et al.*, *Phys. Rev. Lett* **82**, 695 (1999).
4. C. B. Collins *et al.*, *Phys. Rev. C* **61**, 054305 (2000).
5. C. B. Collins *et al.*, *Europhys. Lett.* **57**, 667 (2002).
6. I. Ahmad *et al.*, *Phys. Rev. Lett.* **87**, 072503 (2001).
7. I. Ahmad *et al.*, *Phys. Rev. C* **67**, 041305 (2003).
8. I. Ahmad *et al.*, *Phys. Rev. C* **71**, 024311 (2005).
9. J. J. Carroll *et al.*, *Phys. Lett. B* **679**, 203 (2009).
10. Yang Sun *et al.*, *Phys. Lett. B* **589**, 83 (2004).
11. T. L. Khoo and G. Løvhøiden, *Phys. Lett. B* **67**, 271 (1977).
12. S. M. Mullins *et al.*, *Phys. Lett. B* **393**, 279 (1997);
    S. M. Mullins *et al.*, *Phys. Lett. B* **400**, 401 (1997).
13. A. B. Hayes *et al.*, *Phys. Rev. Lett* **89**, 242501 (2002).
14. F.R. Xu, P.M. Walker, J.A. Sheikh, R. Wyss, *Phys. Lett. B* **435**, 257 (1998).
15. W. Nazarewicz, J. Dudek, R. Bengtsson, T. Bengtsson, I. Ragnarsson, *Nucl. Phys. A* **435**, 397 (1985).
16. J. Dudek, Z. Szymański, T. Werner, *Phys. Rev. C* **23**, 920 (1981).
17. P. Möller and J.R. Nix, *Nucl. Phys. A* **536**, 20 (1992).
18. W. Nazarewicz, M. A. Riley, J. D. Garrett, *Nucl. Phys. A* **512**, 61 (1990).
19. H. C. Pradhan, Y. Nogami, J. Law, *Nucl. Phys. A* **201**, 357 (1973).
20. H. Sakamoto and T. Kishimoto, *Phys. Lett. B* **245**, 321 (1990).
21. W. Satuła and R. Wyss, *Phys. Scr. T* **56**, 159 (1995).
22. F. R. Xu, P. M. Walker, R. Wyss, *Phys. Rev. C* **62**, 014301 (2000).
23. W. D. Myers and W. J. Swiatecki, *Ann. Phys. (N.Y.)* **55**, 395 (1969).
24. V. M. Strutinsky, *Nucl. Phys. A* **95**, 420 (1967).
25. F. R. Xu, P. M. Walker, R. Wyss, *Phys. Rev. C* **59**, 731 (1999).
26. W. Andrejtscheff and P. Petkov, *Phys. Lett. B* **329**, 1 (1994).
27. Ch. Hinke, R. Krücken, R. F. Casten, V. Werner, N. V. Zamfir, *Eur. Phys. J. A* **30**, 357 (2006).
28. C. B. Collins, C. D. Eberhard, J. W. Glesener, J. A. Anderson, *Phys. Rev. C* **37**, 2267 (1988).

STUDY OF NUCLEAR GIANT RESONANCES USING A FERMI-LIQUID METHOD

BAO-XI SUN

Institute of Theoretical Physics, College of Applied Sciences,
Beijing University of Technology, Beijing 100124, China
E-mail: sunbx@bjut.edu.cn

The nuclear giant resonances are studied by using a Fermi-liquid method, and the nuclear collective excitation energies of different values of l are obtained, which are fitted with the centroid energies of the giant resonances of spherical nuclei, respectively. In addition, the relation between the isovector giant resonance and the corresponding isoscalar giant resonance is discussed.

Keywords: Collective excitation; giant resonance; Fermi-liquid method.

In nuclear physics, the Landau parameters are derived microscopically from the ground state energy in the relativistic mean-field approximation, and Landau Fermi liquid theory is used to describe the thermodynamics properties of the nuclear systems, such as the compressibility, the symmetric energy and the hydrodynamics sound velocities.[1] This theory is also studied with an effective chiral Lagrangian to obtain the properties of the nuclear ground state and the link between an effective QCD theory and the nuclear many body theory.[2,3] Moreover, the low momentum nucleon-nucleon potential is applied to calculate the effective interaction between quasi-particles near the Fermi surface, and then the static properties of the nuclear matter are extracted.[4]

Landau Fermi liquid theory is an important method to describe the low energy collective excitation properties of many-body systems, and it can be used to study the problem on giant resonances of nuclei, which is still an hot topic in nuclear physics.[5] With several typical methods, such as the random phase approximation with Skyrme interactions,[6] the relativistic random phase approximation,[7-9] the centroid energies and strength distributions of the giant resonances of nuclei are calculated and compared with the experimental data.

In this work, I will try to calculate the collective excitation energy of the nuclear system within a Fermi-liquid model,[10] which is generalized to the three-dimension situation with the spin of nucleons included.[11] In the calculation, a Lagrangian of Walecka model is used as a microscopic input.[12] The calculation results will be compared with the experimental data of the nuclear giant resonance of some spherical nuclei. Furthermore, I hope to check whether the contribution of Dirac sea of nucleons is essential in the study of collective excitations of nuclear many-body systems.

The liquid equation of motion of the quasi-nucleon in the momentum space can be written as

$$i\frac{\partial}{\partial t}u_\alpha(l,\vec{q},t) = Hu_\alpha(l,\vec{q},t),$$

$$(1)$$

where the Hamiltonian

$$H_{l,l'} = q\left(a_l\delta_{l+1,l'} + a_{l-1}\delta_{l-1,l'}\right)$$
$$\left(v_F^* - \frac{k_F^2}{(2\pi)^3}f_F(l,l)\right)^{1/2}\left(v_F^* - \frac{k_F^2}{(2\pi)^3}f_F(l',l')\right)^{1/2} \quad (2)$$

is hermite with

$$a_l = \sqrt{\frac{(l+1)^2}{(2l+1)(2l+3)}}$$

and v_F^* the Fermi velocity of nucleons in the relativistic mean-field approximation. The eigenvalues of H would give us the frequencies of the collective excitation modes of the nuclear matter. Moreover, it can be seen from Eq. (2) that the exchange interaction between nucleons $f_F(l,l)$ causes the collective excitation of the nuclear matter directly.

The parameters in Ref.[13] are used in the calculation, i.e., $g_\sigma = 10.47$, $g_\omega = 13.80$, $m_\sigma = 520$ MeV, $m_\omega = 783$ MeV and $M_N = 939$ MeV. Since the nucleon near the Fermi surface would be more possible to be excited, we set the value of nucleon momentum $|\vec{q}| = k_F = 1.36$ fm^{-1} in the calculation. When $f_F(l,l) = 0$, the Hamiltonian H has a continuous spectrum and it generates the particle-hole continuum of the nuclear matter in the relativistic mean-field approximation. However, if the value of $f_F(l,l)$ is large enough, in addition to the continuum eigenvalues, the spectrum of H has isolated positive and negative eigenvalues. Actually, the mode with a positive eigenvalue corresponds to the creation of a nuclear collective excitation mode, while the the mode with a negative eigenvalue corresponds to the annihilation of a nuclear collective excitation mode.

The nuclear isoscalar giant resonances actually correspond to the nuclear collective excitations with different values of l. However, the nuclear isovector giant resonances correspond to the nuclear collective excitation states that the collective excitation of protons with the energy $E_S(l)$ is creating , while the collective excitation of neutrons with the energy $E_S(l)$ is annihilating, and vice versa. Hence, the energy of the nuclear isovector giant resonance is about twice of the corresponding isoscalar giant resonance in the nuclear matter, i.e.,

$$E_V(l) = E_S(l) - (-E_S(l)) = 2E_S(l). \tag{3}$$

The experimental data on the nuclear giant resonances in Ref.[14–19] demonstrate that the relation between the energy of nuclear isovector giant resonance and that of nuclear isoscalar giant resonance in Eq. (3) is correct approximately except for the case $l = 1$, which will be discussed in detail specially. In follows, the giant resonance energies with different values of l in the nucleus will be calculated within the Fermi-liquid model.

The nuclear giant monopole mode, $l = 0$, is also called as *breathing mode* of the nucleus. Supposed the proton and neutron densities can be calculated approximately:

$$\rho_p = \rho_0 \frac{Z}{A}, \qquad \rho_n = \rho_0 \frac{N}{A}, \tag{4}$$

the calculated energies for isoscalar and isovector giant monopole resonances of nuclei ^{208}Pb, ^{144}Sm, ^{116}Sn, ^{90}Zr, ^{40}Ca and their corresponding experimental values are listed in Table 1. Since the Fermi momentum of protons $k_F(p)$ is different from that of neutrons, the collective excitation energies of protons and neutrons, $E_0(p)$ and $E_0(n)$, are different from each other. When the effective nucleon mass takes a large value, i.e., $M_N^* = 0.742M_N$, the calculation results of the proton excitation energy for heavy nuclei, such as ^{208}Pb, ^{144}Sm and ^{116}Sn, are fitted with the corresponding experimental centroid energies of the nuclear isoscalar monopole resonance E_{exp}^S, respectively. However, for those light nuclei, ^{90}Zr and ^{40}Ca, we must reduce the effective nucleon mass to $M_N^* = 0.717M_N$, and then the reasonable calculation results fitted with the experimental data are obtained. Moreover, the sum of the excitation energies of protons and neutrons $E_0(p) + E_0(n)$ should correspond to the nuclear isovector giant monopole energy. In Table 1, We can find that calculated values of isovector giant monopole energies for ^{208}Pb, ^{90}Zr and ^{40}Ca are in the range of the experimental errors.

The dipole deformation of the nucleus is really a shift of the center of mass. Thus the isovector giant dipole resonance of the nucleus actually

Table 1. The Fermi momenta and the $l = 0$ collective excitation energies of protons and neutrons for different nuclei with different effective nucleon masses. The corresponding experimental values for the nuclear isoscalar giant monopole resonances E_{exp}^S are taken from Ref.,[15] and the experimental values for the nuclear isovector giant monopole resonances E_{exp}^V from Refs.[17-19]

$l = 0$	^{208}Pb	^{144}Sm	^{116}Sn	^{90}Zr	^{40}Ca
M_N^*/M_N	0.742	0.742	0.742	0.717	0.717
$k_F(p)$ (fm^{-1})	1.26	1.29	1.29	1.31	1.36
$k_F(n)$ (fm^{-1})	1.45	1.42	1.42	1.41	1.36
$E_0(p)$ (MeV)	16.28	15.26	15.26	17.57	15.58
$E_0(n)$ (MeV)	7.05	9.00	9.00	13.13	15.58
$E_0(p) + E_0(n)$ (MeV)	23.33	24.26	24.26	30.7	31.16
E_{exp}^S (MeV)	14.17 ± 0.28	15.39 ± 0.28	16.07 ± 0.12	17.89 ± 0.20	–
E_{exp}^V (MeV)	26.0 ± 3.0	–	–	28.5 ± 2.6	31.1 ± 2.2

corresponds to the creation of the $l = 1$ collective excitation of one kind of nucleons, i.e., protons or neutrons. However, the isoscalar giant dipole resonance in ^{208}Pb with a centroid energy at $E = 22.5$ MeV, which is discovered by studying the (α, α') cross sections at forward angles,[20] should be a compression mode, and corresponds to a creation of the $l = 1$ collective excitation of protons or neutrons and an annihilation of the $l = 1$ collective excitation of the other sort of nucleons, i.e., neutrons or protons simultaneously. The calculation results and the corresponding experimental centroid energies are listed in Table 2. Similarly, we must adjust the effective nucleon mass to obtain the collective excitation energies fitted with the experimental data. It is apparent that in order to obtain a more correct excitation energy, the effective nucleon mass must take a larger value for heavy nuclei, but a smaller value for light nuclei. For the nucleus ^{208}Pb, the summation of the collective excitation energies of protons and neutrons is equal to the experimental energy of the isoscalar giant dipole resonance, approximately.

The calculation results and the corresponding experimental centroid energies of the giant quadrupole resonances of different nuclei are listed in Table 3. The experimental value of the isovector giant quadrupole resonance energy is just twice of the isoscalar giant quadrupole resonance energy for ^{208}Pb, and it manifests the relation between the nuclear isovector giant resonance and the corresponding isoscalar giant resonance in Eq. (3) is correct. Actually, the experimental values for the nuclear giant quadrupole resonance of other nuclei, and even for their monopole giant resonance satisfy the relation in Eq. (3), approximately.

In Table 3, Since the average neutron density is larger than the saturation density of the nuclear matter, the collective excitation energies of the

Table 2. The Fermi momenta and the $l = 1$ collective excitation energies of protons and neutrons for different nuclei with different effective nucleon masses. The corresponding experimental values for the nuclear isoscalar giant dipole resonances E^S_{exp} are taken from Ref.,[20] and the experimental value for the nuclear isovector giant dipole resonance E^V_{exp} from Ref.[16]

$l = 1$	^{208}Pb	^{90}Zr	^{40}Ca
M^*_N/M_N	0.755	0.742	0.70
$k_F(p)$ (fm^{-1})	1.26	1.31	1.36
$k_F(n)$ (fm^{-1})	1.45	1.41	1.36
$E_1(p)$ (MeV)	15.53	15.56	19.58
$E_1(n)$ (MeV)	6.57	11.37	19.58
$E_1(p) + E_1(n)$ (MeV)	22.1	26.93	39.16
E^S_{exp} (MeV)	22.5	–	–
E^V_{exp} (MeV)	13.5 ± 0.2	16.5 ± 0.2	19.8 ± 0.5

nuclei ^{208}Pb and ^{90}Zr for $l = 2$ are less than 10 MeV. These values might correspond to the low-lying excitation states in heavy nuclei, which are in the range of $2 - 6$ MeV for ^{208}Pb.[22]

Table 3. The Fermi momenta and the $l = 2$ collective excitation energies of protons and neutrons for different nuclei. The corresponding experimental values for the nuclear isoscalar giant quadrupole resonances E^S_{exp} are taken from Ref.,[14] and the experimental values for the nuclear isovector giant quadrupole resonances E^V_{exp} from Refs.[7,21]

$l = 2$	^{208}Pb	^{90}Zr	^{40}Ca	^{16}O
M^*_N/M_N	0.742	0.742	0.69	0.69
$k_F(p)$ (fm^{-1})	1.26	1.31	1.36	1.36
$k_F(n)$ (fm^{-1})	1.45	1.41	1.36	1.36
$E_0(p)$ (MeV)	15.02	13.16	18.54	18.54
$E_0(n)$ (MeV)	5.84	8.27	18.54	18.54
$E_0(p) + E_0(n)$ (MeV)	20.86	21.43	37.08	37.08
E^S_{exp} (MeV)	10.9 ± 0.1	14.41 ± 0.1	17.8 ± 0.3	20.7
E^V_{exp} (MeV)	22	–	32.5 ± 1.5	–

The Fermi-liquid model in Ref.[10] is extended to the 3-dimensional Fermion system with the spin taken into account, and then by using the effective Lagrangian of the linear $\sigma - \omega$ model, the Fermi-liquid function is obtained and the isoscalar and isovector giant resonances of some spherical nuclei are studied. we find the centroid energies of the isoscalar giant resonances just correspond to the positive isolated energy levels of the nuclear collective excitation with different values of l, respectively, while the isovector giant resonances except $l = 1$ correspond to the modes that protons(neutrons) are in the creation state of the collective excitation and neutrons(protons) are in the annihilation state of the same l. The low energy excitation of the nuclear giant quadrupole resonance, might be described by the collective excitation of neutrons in the nuclei.

It should be pointed out that the exchange interaction between nucleons plays an important role in the calculation of collective excitations of nuclear systems by using a Fermi-liquid model. Dirac sea of nucleons does not have any influence on the properties of giant resonances of nuclei.

Acknowledgment

This work is supported by the National Natural Science Foundation of China under grant number 10775012.

References

1. T. Matsui, *Nucl. Phys. A* **370**, 365 (1981).
2. B. Friman and M. Rho, *Nucl. Phys. A* **606**, 303 (1996).
3. C. Song, *Phys. Rept.* **347**, 289 (2001).
4. J. W. Holt, G. E. Brown, J. D. Holt et al., *Nucl. Phys. A* **785**, 322 (2007).
5. W. Greiner and J. A. Maruhn, *Nuclear Models*, (Springer-Verlag, Berlin, 1996).
6. I. Hamamoto, H. Sagawa and X. Z. Zhang, *Phys. Rev. C* **57**, R1064 (1998).
7. L. G. Cao, *Ph.D. Dissertation* (2002).
8. Z. Y. Ma, A. Wandelt, N. Van Giai et al., *Nucl. Phys. A* **703**, 222 (2002).
9. J. Daoutidis and P. Ring, *Phys. Rev. C* **80**, 024309 (2009).
10. X. G. Wen, *Quantum Field Theory of Many-Body Systems*, (Oxford University Press, Oxford, 2004).
11. B. X. Sun, *arXiv:1003.1683* [nucl-th].
12. J. D. Walecka, *Theoretical Nuclear and Subnuclear Physics*, 2nd Edition, World Scientific Press, Singapore, 2004.
13. C. J. Horowitz and B. D. Serot, *Nucl. Phys. A* **368**, 503 (1981).
14. A. Van de Woude, *Prog. Part. Nucl. Phys.* **18**, 217 (1987).
15. D. H. Youngblood et al., *Phys. Rev. Lett.* **82**, 691 (1999).
16. B. L. Berman and S. C. Fultz, *Rev. Mod. Phys.* **47**, 713 (1975).
17. A. Erell et al., *Phys. Rev. Lett.* **52**, 2134 (1984).
18. A. Erell et al., *Phys. Rev. C* **34**, 1822 (1986).
19. J. D. Bowman, *Nuclear Structure*, (North-Holland, Amsterdam, 1985). P.549.
20. B. F. Davis et al., *Phys. Rev. Lett.* **79**, 609 (1997).
21. D. M. Drake, K. Aniol, I. Halpern et al., *Phys. Rev. Lett.* **47**, 1581 (1981).
22. P. Ring and P. Schuck, *The Nuclear Many-Body Problem*, (Springer-Verlag, Berlin, 2004).

ROTATIONAL BANDS IN DOUBLY ODD ^{116}Sb

D. P. SUN, S. Y. WANG*, B. QI, C. LIU, C. J. XU, M. Z. GUO, Y. Y. CAI,

L. W. DUAN, G. WANG, X. B. HU, Y. Q. LI, L. LIU

*Shandong Provincial Key Laboratory of Optical Astronomy and Solar-Terrestrial
Environment, School of Space Science and Physics, Shandong University at Weihai,
Weihai 264209, People's Republic of China*
E-mail: sywang@sdu.edu.cn

H. HUA, Z. Y. LI

*School of Physics, and SK Lab. of Nucl. Phys. & Tech., Peking University,
Beijing 100871, People's Republic of China*

L. H. ZHU, X. G. WU, G. S. LI, C. Y. HE, Y. LIU, X. Q. LI, Y. ZHENG,

L. L. WANG, L. WANG

China Institute of Atomic Energy, Beijing 102413, People's Republic of China

High spin states of ^{116}Sb were populated using the ^{114}Cd(^{7}Li, 5n) fusion-evaporation reaction at a beam energy of 48 MeV. The previously reported rotational bands, built on $\pi g_{9/2} \otimes \nu h_{11/2}$ and $\pi g_{9/2} \otimes \nu d_{5/2}$ configurations, have been extended and a new $\Delta I = 1$ band has been identified.

Keywords: High spin; rotational band; configuration.

1. Introduction

The recent experimental studies on high-spin states in ^{108,110,112,114}Sb[1-4] and ^{118}Sb[5] make ^{116}Sb the less studied nucleus among the antimony odd-odd isotopes. In this paper, we report the new experimental results of high spin states in ^{116}Sb.

2. Experiment

High-spin states in ^{116}Sb were populated using the ^{114}Cd(^{7}Li, 5n) reaction. A stack of two self-supporting metallic Cd foils was bombarded with a 48 MeV ^{7}Li beam provided by the HI-13 tandem accelerator at the China Institute of Atomic Energy in Beijing. Gamma-gamma coincidences were mea-

188

sured with 12 Compton-suppressed HPGe-BGO detectors and two LEPS (low-energy photon spectrometer) detectors. Under this condition, approximately 200 million events were recorded.

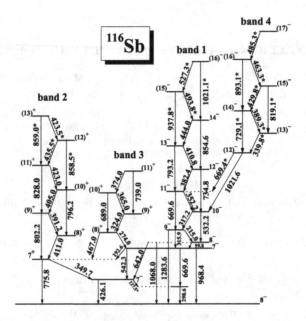

Fig. 1. Rotational band structures of ^{116}Sb proposed in the present work. New transitions observed in the present work are indicated with asterisks.

3. Result and discussion

The level scheme of ^{116}Sb obtained from the present work is shown in Fig. 1. A total of 18 new γ-rays have been added and indicated with asterisks in the level scheme. As shown in Fig. 1, four strongly coupled bands are constructed and labeled as band 1, 2, 3 and 4. Two examples of γ-ray spectra of the rotational bands are presented in Fig. 2.

The experimental alignments and $B(M1)/B(E2)$ ratios for bands 1-4 are extracted and shown in Fig. 3 and 4, respectively. The yrast band (band 1) has been already assigned to the $\pi g_{9/2} \otimes \nu h_{11/2}$ configuration.[6] Bands 2 and 3 were proposed to be built predominantly on the $\pi g_{9/2} \otimes \nu d_{5/2}$ and $\pi g_{9/2} \otimes \nu g_{7/2}$ configurations, respectively.[7] The alignment plot of band 2 exhibits a up-bending at $\hbar\omega = 0.33$ MeV. This phenomenon is caused by

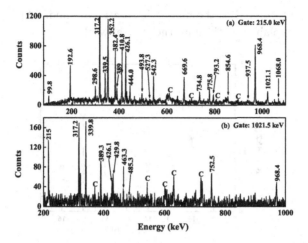

Fig. 2. Examples of γ-γ coincidence spectra in ^{116}Sb.

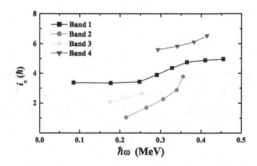

Fig. 3. (Color online) Experimental alignment plots for the bands in ^{116}Sb. The Harris parameters $J_0 = 17\ \hbar^2\mathrm{MeV}^{-1}$ and $J_1 = 12\ \hbar^4\mathrm{MeV}^{-3}$ were used.

the rotational alignment of the first pair of $h_{11/2}$ neutrons.[3] Band 4 is a new band proposed in the present work. The large alignment, as shown in Fig. 3, and the high excitation energy suggest that band 4 is a four quasi-particle band structure. A similar four quasiparticle band structure was also observed in ^{108}Sb.[1] Based on the systematic features of the level schemes, band 4 is assigned to be the $\pi[h_{11/2} \otimes g_{7/2}(g_{9/2})^{-1}] \otimes \nu[g_{7/2}]$ configuration. In addition, band 4 have the strong M1 transitions with relatively weak or absent E2 crossover transitions. A large $B(M1)/B(E2)$ ratios were also observed for band 4 (see Fig. 4). These experimental properties are the characteristics of the magnetic rotational band. Hence, band 4 in ^{116}Sb is

190

tentatively interpreted as the magnetic rotational band. This phenomenon needs to be explained with more theoretical work.

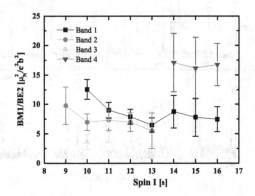

Fig. 4. (Color online) Plot of the $B(M1)/B(E2)$ ratios for bands 1, 2, 3 and 4 as functions of spin.

4. Conclusion

High spin states of odd-odd nucleus ^{116}Sb were populated using the ^{114}Cd(^{7}Li, 5n) reaction at a beam energy of 48 MeV. The level scheme of ^{116}Sb has been extended. A new $\Delta I = 1$ band has been identified, and may be a candidate of magnetic rotational band.

Acknowledgements

This work is supported by the National Natural Science Foundation (Grant Nos. 10875074 and 11005069), the Shandong Natural Science Foundation (Grant No. ZR2010AQ005), and the Major State Research Development Programme (No. 2007CB815005) of China.

References

1. D. G. Jenkins et al., Phys. Rev. C 58 2703(1998).
2. G. J. Lane et al., Phys. Rev. C 55 2127(R) (1997).
3. G. J. Lane et al., Phys. Rev. C 58 127 (1998).
4. E. S. Paul et al., Phys. Rev. C 50 2297 (1994).
5. S. Y. Wang et al., Phys. Rev. C 82 057303 (2010).
6. P. Van Nes et al., Nucl. Phys. A 379 35(1982).
7. M. Fayez-Hassan et al., Nucl. Phys. A 624 401 (1997).

THE STUDY OF THE NEUTRON $N = 90$ NUCLEI

W. X. TENG,* L. R. DAI,** F. PAN, L. LIU and S. H. WANG

Department of Physics, Liaoning Normal University, Dalian 116029, China
E-mail: tengweixin911@yeah.net
**E-mail: dailianrong@gmail.com*

We use a new scheme to describe the $X(5)$ nuclei, in which the $SU(3)$ quadrupole-quadrupole interaction is replaced by the $SO(6)$ cubic $[Q_{(0)} \times Q_{(0)} \times Q_{(0)}]^0$ interaction. Some energy ratios and $B(E2)$ ratios of $X(5)$ nuclei are studied in this work. The results of this new scheme are compared with the corresponding experimental data and those of the traditional $U(5)-SU(3)$ description. It is shown that the results from this new scheme are better than those of the traditional description.

Keywords: Energy levels; $B(E2)$; the $SO(6)$ cubic interaction.

1. Introduction

In the last decades, transitional nuclei were extensively studied in the Interacting Boson Model (IBM).[1-3] Possible descriptions of nuclei at the critical point of shape (phase) transitions from spherical vibrator to axially symmetric rotor, called $X(5)$, have been proposed.[4] In this work, we will study the $X(5)$ nuclei by replacing the well-known $SU(3)$ quadrupole-quadrupole interaction with the $SO(6)$ cubic $[Q_{(0)} \times Q_{(0)} \times Q_{(0)}]^0$ interaction, which was first suggested by van Isacker[5] and further confirmed by Thiamine and Cejinar.[6] We know from Refs.[5,6] that the replacement of the $SU(3)$ quadrupole-quadrupole interaction by the $SO(6)$ cubic $[Q_{(0)} \times Q_{(0)} \times Q_{(0)}]^0$ interaction is an alternative to describe $X(5)$ nuclei in the traditional $U(5)-SU(3)$ transitional region. Experimentally, many nuclei with the $X(5)$ critical point symmetry have been found, such as ^{152}Sm, ^{150}Nd, ^{154}Gd.[7-12] The main purpose of this work is to use this idea to study these three $X(5)$ nuclei.

2. Formulation

The Hamiltonian in this new scheme is

$$\hat{H} = c((1 - \eta)\hat{n}_d + \frac{\eta}{4n}[Q_{(0)} \times Q_{(0)} \times Q_{(0)}]^0), \tag{1}$$

where c is a scaling factor, η is a real number with $0 \leq \eta \leq 1$, $\hat{n}_d$ is d boson operator, n is the number of bosons, and $[Q_{(0)} \times Q_{(0)} \times Q_{(0)}]^0$ is the $SO(6)$ cubic term. This is called UQ scheme. When the cubic term in (1) is replaced by the $SU(3)$ quadrupole-quadrupole interaction which means $U(5)-SU(3)$, the resultant Hamiltonian become the US scheme.

Now we choose the parameters. The parameters c and η in both the UQ and the traditional US schemes are chosen when $\sum_i^{\mathcal{N}} \left| E_{\text{exp}}^i - E_{\text{th}}^i \right|^2 / \mathcal{N}$ reaches the corresponding minimum, where E_{th}^i, and E_{exp}^i are energy of the i-th level calculated, and that of the corresponding experimental value, respectively, and $\mathcal{N}$ is the total number of levels fitted. The quality of fits is estimated by $\sigma = \sqrt{\sum_i^{\mathcal{N}} \left| E_{\text{exp}}^i - E_{th}^i \right|^2 / \mathcal{N}}$. The parameters are listed in Table 1, which shows that the overall fitting quality of the UQ scheme is better than that of the US scheme.

Table 1. The fitting parameters and the corresponding mean-square deviation σ in the US and UQ schemes.

	η	c	σ	η	c	σ
^{150}Nd	0.46	962.4	77.5	0.44	629.0	68.4
^{152}Sm	0.44	869.1	150.8	0.40	598.4	138.0
^{154}Gd	0.42	828.9	91.0	0.36	580.6	95.8

3. Results

Energy ratios are often adopted as benchmarks of the critical point symmetry. Results of these ratios calculated from both the UQ and US schemes and the corresponding experimental values[13–15] are given in Table 2. We can see that the UQ scheme is better than the traditional US scheme if fitting the low-lying spectra of these $X(5)$ nuclei, especially for the ground state band. However, it should be stated that the β band of these nuclei is not well described in the UQ scheme. In particular, the 0_2^+ level fitted in the UQ scheme is lower than the corresponding experimental value, but the overall fitting quality of the UQ scheme is better than that of the US scheme.

The $B(E2)$ branching ratio is another benchmarks of the critical point symmetry, the calculated results of $R_1 \equiv \frac{B(E2;4_1^+ \rightarrow 2_1^+)}{B(E2;2_1^+ \rightarrow 0_1^+)}$, $R_2 \equiv$

Table 2. The experimental and calculated results of energy ratios.

$\dfrac{E(J_i^+)}{E(8_1^+)}$		2_1^+	4_1^+	6_1^+	0_2^+	2_2^+	4_2^+	6_2^+	2_3^+	3_1^+
^{150}Nd	Exp	0.115	0.338	0.638	0.598	0.753	1.007	1.364	0.940	1.063
	US	0.216	0.452	0.715	0.528	0.684	1.005	1.334	0.922	1.062
	UQ	0.113	0.328	0.630	0.386	0.669	0.989	1.332	0.895	1.024
^{152}Sm	Exp	0.108	0.326	0.628	0.608	0.720	0.909	1.165	0.965	1.096
	US	0.224	0.464	0.722	0.522	0.650	0.962	1.279	1.347	1.011
	UQ	0.119	0.333	0.632	0.364	0.665	0.994	1.350	1.224	1.058
^{154}Gd	Exp	0.108	0.324	0.627	0.595	0.713	0.915	1.193	0.871	0.985
	US	0.235	0.447	0.731	0.510	0.621	0.922	1.225	0.853	0.966
	UQ	0.129	0.344	0.638	0.344	0.649	0.985	1.348	0.917	1.073

$\dfrac{B(E2;2_2^+\rightarrow0_1^+)}{B(E2;2_1^+\rightarrow0_1^+)}$, $R_3 \equiv \dfrac{B(E2;2_3^+\rightarrow4_1^+)}{B(E2;2_1^+\rightarrow0_1^+)}$, $R_4 \equiv \dfrac{B(E2;6_1^+\rightarrow4_1^+)}{B(E2;2_1^+\rightarrow0_1^+)}$ for $^{150}N_d$, $^{152}S_m$, and $^{154}G_d$ nuclei are shown in Fig.1. We can see that the $B(E2)$ branching ratios are different in intra-band from those of inter-band transitions. We also see that the results obtained from the UQ scheme are better than those from the US scheme, especially the branching ratios R_2 and R_3 of intra-band transitions obtained from the traditional US scheme are too large in comparison to the corresponding experimental values, while those obtained from the UQ scheme are much closer to the experimental data.[13–15] In addition, the results also show that the UQ scheme is better than the US scheme in fitting both inter-band and intra-band $E2$ transitions.

4. Summary

In summary, the energy ratios and $B(E2)$ ratios of ^{150}Nd, ^{152}Sm, and ^{154}Gd nuclei are studied in the present work, and the results show that the new scheme seems better to describe these $X(5)$ nuclei than the traditional $U(5)-SU(3)$ transitional description.

Acknowledgments

This work was supported by the National Natural Science Foundation of China under Grants Nos. 10975068 and 10775064, Scientific Research Foundation of Liaoning Education Department under Grants Nos. 2009T055 and 2007R28, and Doctoral Fund of Ministry of Education of China (20102136110002).

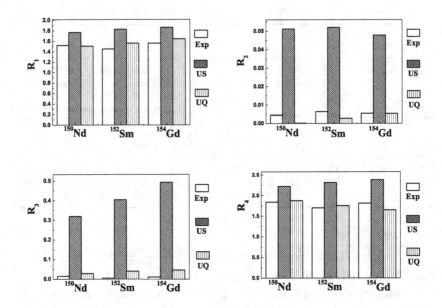

Fig. 1. $B(E2)$ ratios, R_1, R_2, R_3 and R_4 calculated from both the UQ and US schemes and compared with the corresponding experimental values for ^{150}Nd, ^{152}Sm, and ^{154}Gd.

References

1. D. H. Feng, R. Gilmore and S. R. Deans, *Phys. Rev.* **C23**, 1254 (1981).
2. J. Jolie, P. Cejnar, R. F. Casten, S. Heinze, A. Linnemann and V. Werner *Phys. Rev. Lett.* **89**, 182502 (2002).
3. F. Iachello and N. V. Zamfir, *Phys. Rev. Lett.* **92**, 2125014 (2004).
4. F.Iachello, *Phys. Rev. Lett.* **87**, 052502 (2001).
5. P. van. Isacker, *Phys. Rev. Lett.* **83**, 4269 (1999).
6. G. Thiamova, P. Cejnar, *Nucl. Phys.* **A765**, 97 (2006).
7. R. Krücken, B. Albanna, C. Bialik, R. F. Casten, J. R. Cooper, A. Dewald, N. V. Zamfir, C. J. Barton, C. W. Beausang, M. A. Caprio, A. A. Hecht, T. Klug, J. R. Novak, N. Pietralla, and P. Von. Brentano, *Phys. Rev. Lett.* **88**, 232501 (2002).
8. R. F. Casten and N. V. Zamfir, *Phys. Rev. Lett.* **87**, 052503 (2001).
9. D. Tonev, A. Dewald, T. Klug, P. Petkov, J. Jolie, A. Fitzler, O. Möller, S. Heinze, P. von Brentano and R. F. Casten, *Phys. Rev.* **C69**, 034334 (2004).
10. F. Iachello and A. Arima, The Interacting Boson Model (Cambridge University Press, Cambridge) (1987).
11. M. A. Caprio, N. V. Zamfir, R. F. Casten, C. J. Barton, C. W. Beausang, J. R. Cooper, A. A.Hecht, R. Krücken, H. Newman, J. R. Novak, N. Pietralla, A. Wolf and K. E. Zyromski, *Phys. Rev.* **C66**, 054310 (2002).

12. A. Dewald, O. Möller, B. Saha, K. Jessen, A. Fitzler, B. Melon, T. Pissulla, S. Heinze, J. Jolie, K. O. Zell, P. von Brentano, P. Petkov, S. Harissopulos, G. De Angelis, T. Martinez, D. R. Napoli, N. Marginean, M. Axiotis, C. Rusu, D. Tonev, A. Gadea, Y. H. Zhang, D. Bazzacco, S. Lunardi, C. A. Ur, R. Menegazzo and E. Farnea, *J. Phys.* **G31**, S1427 (2005).
13. E. Dermateosian and J. K. Tuli, *Nuclear Data Sheets* **75**, 827 (1995).
14. Artna-Cohen Agda, *Nuclear Data Sheets* **79**, 1 (1996).
15. C. W. Reich, *Nuclear Data Sheets* **110**, 2257 (2009).

DYNAMICAL MODES AND MECHANISMS IN TERNARY REACTION OF ^{197}Au+^{197}Au

JUN-LONG TIAN[1,2,*], HONG-JUN HAO[1], SU-ZHEN YUAN[1] and XUE-QIN LI[3,**]

[1]*School of Physics and Electrical Engineering, Anyang Normal University, Anyang, Henan 455000, China*
[2]*Key Laboratory of Beam Technology and Material Modification of Ministry of Education, Beijing Normal University, Beijing 100875, China*
[3]*National Institute for Radiological Protection, Chinese Center for Disease Control and Prevention, Beijing 100088, China*
E-mail: tianjunlong@gmail.com
**E-mail: lixueqincdc@163.com*

The modes of ternary reaction of ^{197}Au+^{197}Au at an energy of 15A MeV are dynamically studied by the improved quantum molecular dynamics model. Three kinds of modes are found by the time evolution of the configurations of the composite reaction systems: One is the direct mode for which the two time separations of the system happen almost simultaneously. Another is the cascade mode for which a two-step process is clearly shown. The third is oblate mode, a kind of very rare fission event. In this case the composite system deforms to a triangle-like configuration with three necks, and then it forms three equally sized fragments along space-symmetric directions in the reaction plane.

Keywords: ImQMD model; ternary fission modes; direct mode; cascade mode; oblate mode.

Ternary fission of heavy nuclei has been one of the interesting topics in contemporary nuclear physics in the last decades. Early investigations of the dynamics of ternary fission mainly concentrated on fission processes accompanied by light-particle emission, for example, α-particle, Be and C etc. However, for superheavy systems with very massive charges generated by heavy ions reactions, there is very clear evidence for fission into three comparable fragments. The experiment on the ternary partitions of ^{197}Au+^{197}Au at 15 A MeV was carried out by I. Skwira-Chalot et al.[1–3] in 4π geometry using the multidetector array CHIMERA at LNS Catania. The result of the experiment displayed the mass number distributions of fragments according to the mass A_1 (the heaviest), A_2 (the intermediate) and A_3 (the lightest), and the peak of mass number distribution of fragments A_3 was very close to 100. Later, they demonstrate a new reaction mechanism of fast ternary breakup, the colliding system gets torn apart into three massive fragments in which the dominant part of events are nearly aligned along a common re-separation axis.[2,3] It seems to be of great significance to investigate the possible ternary fission modes, since, up to now, little is known about the dynamics of ternary fission for such superheavy systems.

It is worthwhile for us to perform a microscopically dynamic study of the mechanism of the ternary fission. In our previous work, the microscopic transport theory, the improved quantum molecular dynamics (ImQMD) model[4–7] was used to study ternary fission in ^{197}Au+^{197}Au collisions. The calculation results reproduce the characteristic features in ternary breakup events explored in a series of experiments; i.e., the masses and angular distributions of three fragments are comparable in size and the very fast, nearly collinear breakup of the colliding system is dominant in the ternary breakup events.[7–11] Owing to the high charges and high excitation energy in the ^{197}Au+^{197}Au collisions, the mode of fission into nearly equally sized three fragments may become important and the non-equilibrium component of fission is dominant.[12]

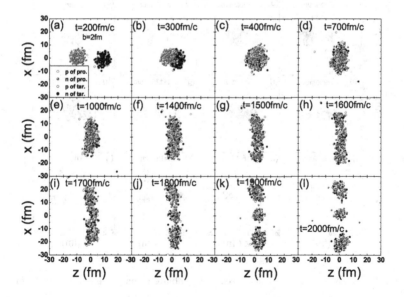

Fig. 1. An example of direct ternary events for ^{197}Au+^{197}Au at 15A MeV with $b = 2$ fm.

198

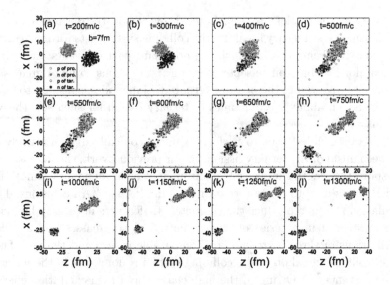

Fig. 2. Same to Fig. 1, but for a cascade ternary event with $b = 7$ fm.

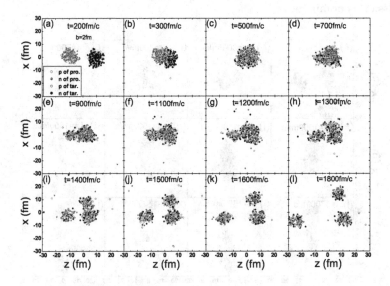

Fig. 3. Same to Fig. 1, but for a sequential oblate ternary event with $b = 2$ fm.

To further understand the new mechanism in a ternary fission of $^{197}\text{Au} + ^{197}\text{Au}$ reactions at 15A MeV, in this talk, we study the modes of the

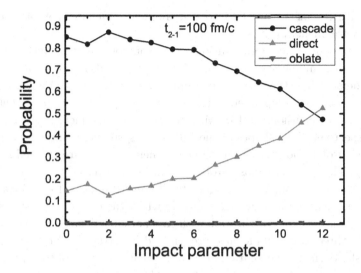

Fig. 4. Impact parameter dependence of the probabilities for direct and cascade ternary events for 197Au + 197Au at 15A MeV. The line with downward triangles denotes oblate ternary events in which three fragments are emitted in a triangular shape. This only happens occasionally.

ternary fission in dynamical process. First, we suppose there are two time separations in a ternary fission event. For convenience, we refer to the first time separation as the first separation and the second time separation as the second separation. The time interval between these two separations is an important quantity used in the exploration of the mechanism of ternary reactions. To phenomenologically clarify the ternary fission modes, as examples, we show snapshots of the time evolution of two typical ternary reaction events for ^{197}Au+^{197}Au at 15A MeV in Figs. 1 and 2, respectively. The spatial configurations and time intervals between these separations are obviously different for these two events. For the event shown in Fig. 1, there exist two preformed necks and the time interval between the two separations is very short (less than 100 fm/c, which is the time step for recording the calculation results); during this interval the two necks break up almost simultaneously, and three fragments are formed from the composite system. This reaction mode is called the direct ternary fission mode. The snapshot shown in Fig. 2 represents a typical cascade ternary fission mode, for which the reaction process can be obviously divided into two steps. In the first step, the reaction system separates into projectile-like and target-like fragments, and in the second step, the projectile-like fragment (or the target-like fragment) breaks into two fragments and the complementary primary fragment survives.

The possible space configurations of the composite system in the ternary process with three mass-comparable fragments are also important features for understanding the ternary dynamics. As examples, in Figs. 1 and 2 we show the time-dependent pictures of two typical ternary events in ^{197}Au+^{197}Au collisions at 15A MeV with impact parameters $b = 2$ and 7 fm, respectively. The shapes of the composite systems for these two events are both prolate elongated. The three fragments formed are emitted almost linearly. These are the most probable cases in the ternary fission process with three mass-comparable fragments. However, in central reactions, say, $b = 2$ fm, we also find very rare events in which the composite systems can take a spherical like configuration with small octuple deformation, in which case the three fragments formed are emitted in a triangular configuration. Fig. 3 shows the time evolution of the shape configuration of the system for this typical event. One sees that at $t = 500$ fm/c the composite system with a roughly spherical shape is formed and then develops into a pear shape at 700 fm/c. At $t = 1100$ fm/c, one neck is formed. At about 1400 fm/c, the first time break takes place and the composite system divides into two parts. The larger part undergoes further elongation and eventually separates into two fragments again at 1600 fm/c. The lifetime of the composite system in these rare events is much longer than that in other cases. Although this kind of ternary events is very rare and only happens in central reactions, the special configuration and quite long lifetime of the formed composite systems are very interesting for some purposes, for example, for studies of exotic nuclear structure and spontaneous emission of positrons in strong electromagnetic fields.

Figure 4 shows the impact parameter dependence of the probabilities for the direct, cascade ternary fission events for the reaction ^{197}Au+^{197}Au at 15 A MeV. From Fig. 4, one can see that the probability for direct ternary reactions increases, and that for cascade ternary reactions decreases with impact parameters, and at very large impact parameters(peripheral reactions) the probability for direct fission events excesses that for cascade. It means that the ternary fission reaction is dominated by the binary breakup or binary process with simultaneously emitted light-charged particles at the neck for the events at peripheral reactions. So, the increase of direct mode with impact parameter means that, in this case, the probability for a binary process with a neck emission increases with impact parameter.

In summary, the modes of ternary fission for the very heavy system

^{197}Au+^{197}Au at energy of 15 A MeV has been investigated by using the ImQMD model. It shows that three kinds of modes exist by the time dependent snapshots of typical ternary events. One is the direct mode for which the two time separations of the system happen almost simultaneously (the time interval between two separations being much less than 100 fm/c). Another is the cascade mode for which a two-step process is clearly shown. The third is oblate mode, a kind of very rare fission event. In the case the composite system deforms to a triangle-like configuration with three necks, and then it forms three equally sized fragments along space-symmetric directions in the reaction plane. The time evolution of the configurations of the composite reaction systems is also studied. We find that for most of the ternary fission events with the features found in the experiments, the configuration of the composite system has two-preformed-neck shape. Finally, the impact parameter dependence of the probabilities for the three ternary fission modes are calculated. It is found that the probability for a binary process with a neck emission increases with impact parameter.

Acknowledgments

One of the authors (J.L. Tian) is grateful to Profs. X.Z. Wu and Z.X. Li for fruitful discussions. This work is supported by the National Natural Science Foundation of China (Nos. 11005003, 10975095, 11005002 and 11047108), the Natural Science Foundation of He'nan Educational Committee (Nos. 2011A140001, 2010B140001).

References

1. I. S. Chalot, K. S. Wilczynska, J. Wilczynski, et al., Int. J. Mod. Phys. E **15**, 495 (2006); **16**, 511 (2007); **17**, 41 (2008).
2. I. S. Chalot, K. S. Wilczynska, J. Wilczynski, et al., Phys. Rev. Lett. **101**, 262701 (2008).
3. J. Wilczynski, I. S. Chalot, K. S. Wilczynska, et al., Phys. Rev. C **81**, 067604 (2010).
4. N. Wang, Z. X. Li and X. Z. Wu, Phys. Rev. C **65**, 064608 (2002).
5. N. Wang, Z. X. Li, X. Z. Wu, et al., Phys. Rev. C **69**, 034608 (2004).
6. N. Wang, Z. X. Li, X. Z. Wu, et al., Mod. Phys. Lett. A **20**, 2619 (2005).
7. J. L. Tian, X. Z. Wu, Z. X. Li, et al., Phys. Rev. C **77**, 064603 (2008).
8. J. L. Tian, X. Li, X. Z. Wu, et al., Chin. Phys. C **SI**, 109 (2009).
9. J. L. Tian, X. Li, X. Z. Wu, et al., Chin. Phys. Lett. **26**, 062502 (2009).
10. J. L. Tian, X. Li, X. Z. Wu, et al., Chin. Phys. Lett. **26**, 082504 (2009).
11. J. L. Tian, X. Z. Wu, Z. X. Li, et al., Phys. Rev. C **82**, 054608 (2010).
12. J. L. Tian, X. Li, S. W. Yan, et al., Int. J. Mod. Phys. E **19**, 307 (2010).

DYNAMICAL STUDY OF $X(3872)$ AS A $D^0 \bar{D}^{*0}$ MOLECULAR STATE

B. WANG, H. K. YUAN and H. CHEN*

*School of Physical Science and Technology, Southwest University,
Chongqing, 400715, China*
E-mail: chenh@swu.edu.cn
www.swu.edu.cn

The possibility is considered that $X(3872)$ may be a bound state of $D^0 \bar{D}^{*0}$ or $\bar{D}^0 D^{*0}$. By assuming that the short range force between mesons is governed by one-glue-mediated quark exchange between two mesons and long range force by sigma and pion meson exchange, we calculate the binding energy of $D^0 \bar{D}^{*0}$ system by solving the Schrödinger equation. Results indicate that there exists a weakly bound S-wave $D^0 \bar{D}^{*0}$ state and the binding energy is less sensitive to the parameters of the meson exchange potential.

Keywords: Molecular state; quark exchange; one sigma exchange; one pion exchange.

1. Introduction

The narrow charmonium-like state $X(3872)$ was discovered in $B^{\pm} \rightarrow K^{\pm}\pi^+\pi^- J/\psi$ decays in 2003.[1] It is very hard to consider $X(3872)$ as a charmonium state in the context of the conventional quark model, and moreover, a charmonium state with isospin $I = 0$ does not easily decay into $J/\psi\rho$. Because $X(3872)$ lies close to the threshold of $D^0 \bar{D}^{*0}$, it is popular to assign it as a weakly bound hadronic molecule.[2-7] As we know, there is still an open problem about the mechanism of the short distance quark-quark interaction, whether one gluon exchange (OGE) or vector meson exchange is dominant or whether both of them are needed.

In this paper, we will study $X(3872)$ as a $D^0 \bar{D}^{*0}$ ($\bar{D}^0 D^{*0}$) molecular state. We assume that the short distance interaction is governed by OGE between two mesons as in the original work,[4,8] while the long range dynamics is mediated by one pion and sigma exchange (OPE and OSE). Then we calculate the binding energy of $D^0 \bar{D}^{*0}$ and find that it is easier to form a $D^0\bar{D}^{*0}$ S-wave bound state with the potential derived from OGE, OPE and OSE, and that these results are less sensitive to the parameters.

2. Quark exchange induced interaction

First of all, the meson-meson potential induced by quark exchanges is assumed to be the sum of OGE and confining potentials.[4,8] The meson-meson reaction is described by quark-quark scattering process, which leads to the quark rearrangement between the two singlet clusters and the subsequent formation of the final two meson. The extracted meson-meson potential from the quark exchange process can be found in Ref.[8] The spin-spin hyperfine (SS), color coulomb (CC) and linear confinement (LC) part are

$$V_{D\bar{D}^*}^{(I=0)}(r)_{SS} = -\frac{2^{3/2}}{27\pi^{1/2}}\frac{\alpha_s\beta^3}{m^2}e^{-\beta^2r^2/2}, \tag{1}$$

$$V_{D\bar{D}^*}^{(I=0)}(r)_{CC} = \frac{4\alpha_s}{9r}\left[1 + (2/\pi)^{1/2}\beta r - 4\text{Erf}(\beta r/2)\right]e^{-\beta^2r^2/2}, \tag{2}$$

$$V_{D\bar{D}^*}^{(I=0)}(r)_{LC} = -\frac{2b}{6\beta}\left[\beta re^{-\beta^2r^2/2} + \frac{2^{3/2}e^{-\beta^2r^2/2}}{\pi^{1/2}} - \left(\beta r + \frac{2}{\beta r}\right)\right.$$
$$\left. \times\text{Erf}(\beta r/2)e^{-\beta^2r^2/2} - \frac{2}{\pi^{1/2}}e^{-3\beta^2r^2/4}\right], \tag{3}$$

where Erf(r) denotes the error function, b is the string tension, m is the up (down) quark mass and α_s is the coupling constant. The parameters were $\alpha_s = 0.5$,[9] $b = 0.21$ GeV^2 and 0.33, 0.55, and 1.6 GeV for light, strange, and charm quark masses respectively.[8] Regarding the harmonic oscillator parameter β of the charmed meson, its value is 0.468 or 0.422 for different potential.[10] Using the expression between f_D[11] and β,[12] we also educe $\beta_D = 0.35$. Here we adopt the moderate values $\beta = 0.405$. With these parameters, the potential induced by quark exchange is shown in Fig. 1(a).

3. OPE and OSE potentials

The long range OSE and OPE interactions, which are expected to dominate the physics of a weakly bound state, occur between two mesons. Therefore, we append OPE and OSE to the nonrelativistic quark model. One boson exchange potential (marked by OBEP) induced by OPE and OSE is,[5]

$$V_{OBEP}(\mathbf{r}) = -g_\sigma^2 Y_\sigma(\mathbf{r}) + \frac{g^2}{6f_\pi^2}Y_\pi(\mathbf{r}) \tag{4}$$

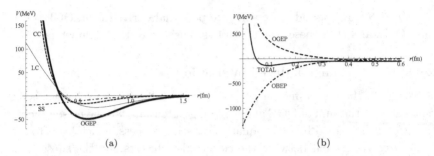

(a)   (b)

Fig. 1. (a) OGE and linear confining potential. The dot-dashed and dashed lines stand for color coulomb and linear confining potential respectively. The thin and thick solid lines correspond to spin-spin hyperfine potential and total effect (marked by OGEP) of the three kinds of potential respectively. (b) The meson-meson potential with $g = 0.66$ and $\Lambda = 2$ GeV. The dashed line represents OGEP. The dot-dashed line represents OBEP. The solid line corresponds to the total potential of OGEP and OBEP.

Taking into account the monopole form factor[13] $F(q) = (\Lambda^2 - M^2)/(\Lambda^2 - q^2)$ (Λ is a cutoff parameter), one can take the following regulated forms:[7]

$$Y_\sigma(r) = \frac{1}{4\pi r}(e^{-M_\sigma r} - e^{-\Lambda r}) - \frac{\eta'^2}{8\pi\Lambda}e^{-\Lambda r}, \qquad (5)$$

$$Y_\pi(r) = -\frac{\mu^2}{4\pi r}[\cos(\mu r) - e^{-\delta r}] - \frac{\eta^2\delta}{8\pi}e^{-\delta r}, \qquad (6)$$

where $\eta = \sqrt{\Lambda^2 - M_\pi^2}$, $\eta' = \sqrt{\Lambda^2 - M_\sigma^2}$ and $\delta = \sqrt{\Lambda^2 - q_0^2}$. In Fig.1(b), the parameters were $g_\sigma = 0.76$,[7] $M_\sigma = 0.6$GeV, $M_{\pi^0} = 0.135$GeV, $M_{D^0} = 1.8648$GeV, $M_{D^{*0}} = 2.007$GeV.[14]

4. Numerical results

Since the mass of the $D^0\bar{D}^{*0}$ system is heavy, a nonrelativistic approximation is meaningful. Then we solve the Schrödinger equation by a program[15] for finding a bound state, which has at least one negative eigenvalue. In Table 1, the symbol '+' denotes that there is no negative eigenvalue. We can see that the binding energy is sensitive to the parameters of g and Λ considering only OPEP or OBEP. While OGEP and OBEP are included, the solutions are less sensitive to the parameters. And the larger the parameters becomes, the easier one gets the negative eigenvalue. Furthermore, the larger the absolute value of binding energy E_b is, the smaller r_{rms} is. Otherwise, the typical Λ is expected to be around 1–2 GeV. And it is reasonable to adopt $\Lambda = 2$ and $g = 0.66$.[16] Under this condition, combining the effect of OGEP and OBEP, we can explain $X(3872)$ as a loose S-wave

Table 1. The binding energy and the rms radius with several kinds of potential. The units of binding energy E_b and the root-mean-square(rms) radius r_{rms} are MeV and fm respectively.

Parameters		OPEP		OBEP		OBEP+OGEP	
g	Λ(GeV)	E_b	r_{rms}	E_b	r_{rms}	E_b	r_{rms}
	3	+	+	+	+	-0.04	11.11
0.59	5	+	+	+	+	-0.54	4.41
	6	-11.24	0.98	-27.35	0.66	-2.62	2.37
	1	+	+	+	+	-4.77×10^{-3}	9.91
0.66	2	+	+	+	+	-0.11	9.09
	3	+	+	+	+	-0.31	5.82
	1	+	+	+	+	-1.25	3.29
1	2	-0.20	6.50	-1.41	2.85	-6.80	1.58
	3	-169.10	0.32	-187.53	0.31	-69.22	0.54

molecular state with a binding energy -0.11 MeV and $r_{rms} = 9.09$ fm, which is consistent with the prediction $E_b \geq -0.4$ MeV and $r_{rms} \geq 7$ fm.[3]

5. Summary

In this work, we investigate X(3872) as a $D^0 \bar{D}^{*0}$ ($\bar{D}^0 D^{*0}$) molecular state bound by including the short-range quark exchange induced interactions and long-range meson exchange interactions. The results indicate that the negative eigenvalue is less sensitive to the parameters g and Λ with above interactions. Combining the effect of OGEP and OBEP with the reasonable parameters, we can explain $X(3872)$ as a very loose molecular state.

References

1. S. K. Choiet et al. (Belle Collaboration), Phys. Rev. Lett. **91**, 262001 (2003).
2. M. B. Voloshin and L. B. Okun, JETP Lett. **23**, 333 (1976).
3. F. E. Close and P. R. Page, Phys. Lett. B **578**, 119 (2004).
4. E. S. Swanson, Phys. Lett. B **588**, 189 (2004).
5. X. Liu, Z. G. Luo, Y. R. Liu and S. L. Zhu, Eur. Phys. J. C **61**, 411 (2009).
6. Y. R. Liu and Z. Y. Zhang, Phys. Rev. C **79**, 035206 (2009).
7. Y. R. Liu, X. Liu, W. Z. Deng and S. L. Zhu, Eur. Phys. J. C **56**, 63 (2008).
8. T. Barnes et al., Phys. Rev. C **60**, 045202 (1999).
9. D. Scora and N. Isgur, Phys. Rev. D **52**, 2783 (1995).
10. H. M. Choi, Phys. Rev. D **75**, 073016 (2007).
11. M. Artuso et al. (CLEO Collaboration), Phys. Rev. Lett. **95**, 251801 (2005).
12. H. Chen and R. G. Ping, J. Phys. G **34**, 2679 (2007).
13. N. A. Törnqvist, Phys. Rev. Lett. **67**, 556 (1991); Z. Phys. C **61**, 525 (1994).
14. C. Amsler et al. (Particle Data Group), Phys. Lett. B **667**, 1 (2008).
15. W. Lucha and F. F. Schoberl, Int. J. Mod. Phys. C **10**, 607 (1999).
16. S. Ahmed et al. (CLEO Collaboration), Phys. Rev. Lett. **87**, 251801 (2001).

SUPER-HEAVY STABILITY ISLAND WITH A SEMI-EMPIRICAL NUCLEAR MASS FORMULA

N. WANG* and M. LIU

Department of Physics, Guangxi Normal University
Guilin 541004, P. R. China
*E-mail: wangning@gxnu.edu.cn

Based on a semi-empirical nuclear mass formula, the super-heavy stability island is investigated. From the calculated shell corrections of super-heavy nuclei, the region $N = 172 - 178$, $Z = 116 - 120$ with shell corrections about -6 MeV roughly gives the position of the super-heavy stability island. The probability to synthesize nuclei with $Z = 126$ may be much smaller than that of produced super-heavy nuclei already, according to the obtained shell corrections and the proton drip line.

Keywords: Super-heavy nuclei; nuclear mass formula; shell correction.

1. Introduction

The synthesis of super-heavy nuclei has been studied for about half century and great achievements have been obtained.[1-3] However the central position of the super-heavy stability island is still uncertain. The predicted proton magic number could be $Z = 114$, 120 or 126, and the neutron number could be $N = 172$, 178 or 184, based on different mean field models and different model parameters.[4-6] Careful calculations of the shell corrections and the binding energies for super-heavy nuclei play a key role for determination of the central position of the stability island.

In Refs.,[6,7] we proposed a semi-empirical nuclear mass formula based on the macroscopic-microscopic method[5] together with the Skyrme energy density functional. To extend the mass formula to super-heavy nuclei and the nuclei far from the β-stability line, we pay a special attention to study the isospin and mass dependence of the model parameters including symmetry energy coefficient and the symmetry potential. In addition, we consider the mirror nuclei constraint due to isospin symmetry in nuclear physics, which improves the precision of mass calculation significantly. In this talk, we first briefly introduce the nuclear mass formula. Then, based on the calculations from the model, we investigate the shell corrections of super-heavy nuclei and the stability island.

2. An improved macroscopic-microscopic mass formula

In the proposed model, the total energy of a nucleus is written as a sum of the liquid-drop energy and the Strutinsky shell correction ΔE,

$$E(A, Z, \beta) = E_{\mathrm{LD}}(A, Z) \prod_{k \geq 2} \left(1 + b_k \beta_k^2\right) + \Delta E(A, Z, \beta). \qquad (1)$$

The liquid drop energy of a spherical nucleus $E_{\mathrm{LD}}(A, Z)$ is described by a modified Bethe-Weizsäcker mass formula,

$$E_{\mathrm{LD}}(A, Z) = a_v A + a_s A^{2/3} + E_C + a_{\mathrm{sym}} I^2 A + a_{\mathrm{pair}} A^{-1/3} \delta_{np} \qquad (2)$$

with isospin asymmetry $I = (N - Z)/A$, the Coulomb term

$$E_C = a_c \frac{Z^2}{A^{1/3}} [1 - Z^{-2/3}] \qquad (3)$$

and the symmetry energy coefficient,

$$a_{\mathrm{sym}} = c_{\mathrm{sym}} \left[1 - \frac{\kappa}{A^{1/3}} + \frac{2 - |I|}{2 + |I|A}\right]. \qquad (4)$$

Here, we introduce an isospin-dependent term in the symmetry energy coefficient for a description of the Wigner term (some details will be discussed later). The a_{pair} term empirically describes the pairing effect (see Ref.[6] for details). The terms with b_k describe the contribution of nuclear deformation to the macroscopic energy, and the mass dependence of b_k is written as,

$$b_k = \left(\frac{k}{2}\right) g_1 A^{1/3} + \left(\frac{k}{2}\right)^2 g_2 A^{-1/3}, \qquad (5)$$

which greatly reduces the computation time for the calculation of deformed nuclei.

The microscopic shell correction

$$\Delta E = c_1 E_{\mathrm{sh}} + |I| E'_{\mathrm{sh}} \qquad (6)$$

is obtained with the traditional Strutinsky procedure by setting the order $p = 6$ of the Gauss-Hermite polynomials and the smoothing parameter

$\gamma = 1.2\hbar\omega_0$ with $\hbar\omega_0 = 41A^{-1/3}$ MeV. E_{sh} and E'_{sh} denote the shell energy of a nucleus and of its mirror nucleus, respectively. The additionally introduced $|I|E'_{\text{sh}}$ term is to empirically take into account the mirror nuclei constraint ($\Delta E - \Delta E' \approx 0$, that is to say, a small value for the shell correction difference of a nucleus and its mirror nucleus due to the charge-symmetric and charge-independent nuclear force), with which the rms deviation of masses can be considerably reduced by about 10%. The single-particle levels are obtained under an axially deformed Woods-Saxon potential[8] in which the depth V_q of the central potential ($q = p$ for protons and $q = n$ for neutrons) is written as

$$V_q = V_0 \pm V_s I \tag{7}$$

with the plus sign for neutrons and the minus sign for protons. V_s is the isospin-asymmetric part of the potential depth. We set the symmetry potential $V_s = a_{\text{sym}}$. Simultaneously, the isospin-dependent spin-orbit strength is adopted based on the Skyrme energy density functional,

$$\lambda = \lambda_0 \left(1 + \frac{N_i}{A}\right) \tag{8}$$

with $N_i = Z$ for protons and $N_i = N$ for neutrons, which strongly affects the shell structure of neutron-rich nuclei and super-heavy nuclei. In addition, we assume and set the radius $R = r_0 A^{1/3}$ and surface diffuseness a of the single particle potential of protons equal to those of neutrons for simplicity. For protons the Coulomb potential is additionally involved.

In this model, the isospin effects in both macroscopic and microscopic part of the formula are self-consistently considered, with which the number of model parameters is considerably reduced compared with the finite range droplet model. Here, we have 13 independent parameters a_v, a_s, a_c, c_{sym}, κ, a_{pair}, g_1, g_2, c_1, V_0, r_0, a, λ_0 in the nuclear mass model. Based on the 2149 measured nuclear masses,[9] the optimal model parameters (WS*) are obtained and listed in Table 1.

3. Details of the model and some results

In this section, we will first introduce the Wigner term in the symmetry energy coefficient. Then, we will present some calculated results of nuclear masses and shell corrections. Finally, some properties of super-heavy nuclei are investigated with the proposed model.

Table 1. Model parameters.

parameter	WS*
a_v (MeV)	−15.6223
a_s (MeV)	18.0571
a_c (MeV)	0.7194
c_{sym}(MeV)	29.1563
κ	1.3484
a_{pair}(MeV)	−5.4423
g_1	0.00895
g_2	−0.4632
c_1	0.6297
V_0 (MeV)	−46.8784
r_0 (fm)	1.3840
a (fm)	0.7842
λ_0	26.3163

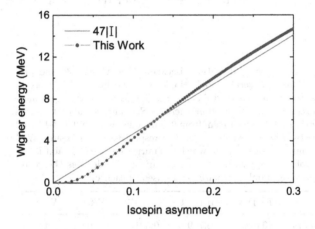

Fig. 1. (Color online) Wigner energies of nuclei along the beta-stability line. The filled circles and the straight line denote the results of this work and those of Satula et al.,[10] respectively.

It is known that the isospin effect plays a key role for neutron-rich nuclei and super-heavy nuclei. The nuclear binding energies, when plotted along isobaric sequences that cross the $N = Z$ locus, exhibit a slope discontinuity roughly proportional to $|I|$, which has been interpreted in terms of isospin symmetry or Wigner $SU(4)$ symmetry and is usually referred to as the Wigner effect. In this work, the Wigner effect is incorporated in the symmetry energy coefficient. The introduced I term in a_{sym} roughly leads to a correction E_W to the binding energy of the nucleus,

$$E_W = c_{\text{sym}} I^2 A \left[\frac{2 - |I|}{2 + |I|A} \right] \approx 2 c_{\text{sym}} |I| - c_{\text{sym}} |I|^2 + ..., \qquad (9)$$

which is known as the Wigner term. In Fig. 1, we show the comparison of the Wigner energies E_W of nuclei along the beta-stability line calculated with different models as a function of isospin asymmetry I. The straight line denotes the results of traditional Wigner energy about $47|I|$ MeV in Ref.[10] The filled circles denote the results of this work, which is close to the results from traditional method. Compared with the case without the I term being taken into account, the rms deviation of 2149 nuclear masses can be reduced by 6%. Furthermore, when the isospin dependence of symmetry energy coefficient is taken into account, the obtained optimal c_{sym} changes from 26 to 29 MeV which is close to the calculated symmetry energy coefficient of nuclear matter at saturation density from the Skyrme energy density functional.[6]

Table 2. rms σ deviations between data AME2003[9] and predictions of several models (in MeV). The line $\sigma(M)$ refers to all the 2149 measured masses, the line $\sigma(S_n)$ to the 1988 measured neutron separation energies S_n. The calculated masses with FRDM are taken from.[5] The masses with HFB-14 and HFB-17 are taken from[11] and,[12] respectively. The results WS[6] in our previous work in which the mirror nuclei constraint is not involved, are also listed for comparison. N_p denotes the corresponding number of parameters used in each model.

	FRDM	HFB-14	HFB-17	WS	WS*
$\sigma(M)$	0.656	0.729	0.581	0.516	0.441
$\sigma(S_n)$	0.399	0.598	0.506	0.346	0.332
N_p	31	24	24	15	15

With the obtained optimal parameters of mass formula listed in Table 1, the rms deviations of the 2149 nuclear masses is significantly reduced to 0.441 MeV and the rms deviation of the neutron separation energies of 1988 nuclei is reduced to 0.332 MeV (see Table 2). Compared with the FRDM, the rms error for the 2149 nuclear masses is considerably reduced with WS*, from 0.656 to 0.441 MeV, whilst the number of

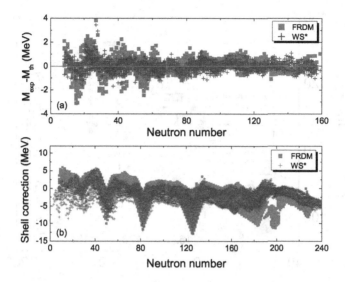

Fig. 2. (Color online) (a) Deviations between the calculated nuclear masses from the experimental data. (b) Calculated shell corrections ΔE of nuclei (crosses). The squares denote the microscopic energy of nuclei with the FRDM model (column E_{mic} of the table of Ref.[5]).

parameters in the model is reduced by a factor of two. Fig. 2(a) shows the deviations between the calculated nuclear masses in this work from the experimental data. The precision of calculated masses is obviously improved in WS*, especially for light and intermediate nuclei. In Fig. 2(b), we show the calculated shell corrections ΔE of nuclei with our model and the microscopic energy (mainly including the shell correction and the deformation energy) obtained in the finite-range droplet model. For intermediate and known heavy nuclei, the results of the two approaches are comparable and both of them reproduce the known magic numbers very well. The deviations are large for light nuclei and super-heavy nuclei. The proposed model can remarkably well reproduce the shell gaps and alpha-decay energies of synthesized super-heavy nuclei (the rms deviation of the α-decay energies of 46 super-heavy nuclei is reduced to 0.263 MeV).[7] These results give us considerable confidence to explore the super-heavy stability island.

The magic numbers 2, 8, 20, 28, 50, 82 and 126 (for neutron) are well determined from the spherical shapes of nuclei and the discontinuities of the neutron separation energy, etc. for most measured nuclei. For super-heavy

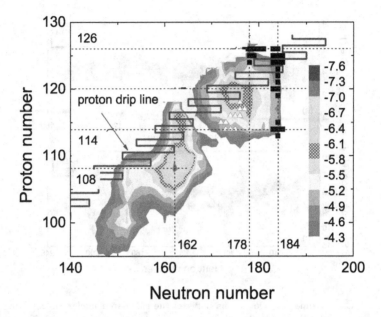

Fig. 3. (Color online) Shell correction energies of nuclei in super-heavy region from WS* calculations. The black squares denote the nearly spherical nuclei (calculated $|\beta_2| \leq$ 0.01) and the triangles denote the synthesized super-heavy nuclei in the "hot" fusion reactions.[2,3] The dark gray zigzag line denotes the calculated proton drip line. The dashed lines give the possible magic number in super-heavy region.

nuclei, the determination of magic number however becomes a little complicated. In Fig. 3, we show the shell corrections of super-heavy nuclei. The black squares denote the nearly spherical nuclei (calculated $|\beta_2| \leq 0.01$). If to determine the magic number from the shapes in spherical of nuclei, one can see that neutron number $N = 184$ is an obvious magic number. However, the largest shell corrections of nuclei in super-heavy region locate around $N = 162$ and 178. The results of WS* and FRDM indicate that ^{270}Hs ($N = 162$ and $Z = 108$) is a deformed doubly-magic nucleus from the large shell correction in absolute value. The shades in Fig. 3 show the region ($N = 172 - 178$, $Z = 116 - 120$) of nuclei with shell corrections of about -6 MeV, which roughly gives the boundaries of the super-heavy island based on the calculated shell correction of nuclei with WS*. In addition, one can see from the calculated proton drip line that nuclei with $Z = 126$ and $N \leq 184$ locate beyond the proton drip line and the corresponding shell corrections (in absolute value) are much smaller than those of known super-heavy nuclei, which indicates that the probability to synthesize nuclei with $Z = 126$ would be much smaller than that of produced super-heavy nuclei already.

4. Summary

Based on our proposed nuclear mass formula, we investigated the super-heavy stability island. According to the calculated masses, the rms deviation with respect to 2149 measured nuclear masses is reduced to 0.441 MeV and the rms deviation of the neutron separation energies of 1988 nuclei falls to 0.332 MeV, which give us considerable confidence to study the properties of super-heavy nuclei. From the calculated shell corrections of super-heavy nuclei, the region $N = 172 - 178$, $Z = 116 - 120$ with shell corrections about -6 MeV roughly gives the position of the super-heavy stability island. The probability to synthesize nuclei with $Z = 126$ may be much smaller than that of produced super-heavy nuclei already, since nuclei with $Z = 126$ and $N \leq 184$ locate beyond the proton drip line and the corresponding shell corrections (in absolute value) are much smaller than those of known super-heavy nuclei. To improve the reliability of the predication on super-heavy nuclei, one needs to further improve the nuclear mass model. Our preliminary results show that the rms deviation with respect to 2149 measured nuclear masses can be reduced to about 340 keV after some residual corrections are considered in the model, and the obtained position of the super-heavy stability island is generally unchanged.

Acknowledgments

This work was supported by National Natural Science Foundation of China, Nos 10875031, 10847004 and 10979024. The FORTRAN program of the proposed mass formula and the corresponding mass tables are available at http://www.imqmd.com/.

References

1. S. Hofmann and G. Mnzenberg, *Rev. Mod. Phys.* **72**, 733 (2000).
2. Yu. Ts. Oganessian, V. K. Utyonkov *et al.*, *Phys. Rev. C* **69** (2004) 054607.
3. Yu. Ts. Oganessian, V. K. Utyonkov *et al.*, *Phys. Rev. C* **74** (2006) 044602.
4. S. Ćwiok, J. Dobaczewski, P.-H. Heenen, P. Magierski, W. Nazarewicz, *Nucl. Phys. A* **611**, 211 (1996).
5. P. Möller, J. R. Nix *et al.*, *At. Data and Nucl. Data Tables* **59** (1995) 185.
6. Ning Wang, Min Liu and Xizhen Wu, *Phys. Rev. C* **81**, 044322 (2010).

214

7. Ning Wang, Zuoying Liang, Min Liu and Xizhen Wu, *Phys. Rev. C* **82**, 044304 (2010).
8. S. Cwoik, J. Dudek *et al.*, *Comp. Phys. Comm.* **46** (1987) 379.
9. G. Audi, A. H. Wapstra and C. Thibault, *Nucl. Phys. A* **729**, 337 (2003).
10. W. Satula , D. J. Dean *et al.*, *Phys. Lett. B* **407** (1997) 103.
11. S. Goriely, M. Samyn and J. M. Pearson, *Phys. Rev. C* **75** (2007) 064312.
12. S. Goriely, N. Chamel and J. M. Pearson, *Phys. Rev. Lett.* **102** (2009) 152503.

PSEUDOSPIN PARTNER BANDS IN ^{118}Sb

S. Y. WANG*, B. QI, D. P. SUN, X. L. REN, B. T. DUAN, F. CHENG, C. LIU,

C. J. XU, and L. LIU

Shandong Provincial Key Laboratory of Optical Astronomy and Solar-Terrestrial Environment, School of Space Science and Physics, Shandong University at Weihai, Weihai 264209, People's Republic of China
E-mail: sywang@sdu.edu.cn

H. HUA, and Z. Y. LI

School of Physics, and SK Lab. of Nucl. Phys. & Tech., Peking University, Beijing 100871, People's Republic of China

J. M. YAO

School of Physical Science and Technology, Southwest University, Chongqing 400715, People's Republic of China

L. H. ZHU, X. G. WU, G. S. LI, C. Y. HE, Y. LIU, X. Q. LI, Y. ZHENG,

L. L. WANG, and L. WANG

China Institute of Atomic Energy, Beijing 102413, People's Republic of China

High-spin states of ^{118}Sb have been studied by in-beam γ-spectroscopy following the reaction ^{116}Cd(^{7}Li, 5n)^{118}Sb. The previously known band structures have been extended. One new rotational band has been identified, and assigned the $\pi g_{9/2}^{-1} \otimes \nu g_{7/2}$ configuration. Two positive parity bands observed in the present work are suggested as a pair of pseudospin partner bands.

Keywords: High spin states; configuration; pseudospin partner bands.

1. Introduction

The concept of pseudospin symmetry has been introduced in nuclear structure physics many years ago,[1] in order to account for the quasi-degeneracy between two single-particle orbitals. It has been attributed to an accidental cancelation between the spin-orbit field $\vec{l} \cdot \vec{s}$ and the $\vec{l}^2$ term in the Nilsson potential. By introducing the pseudo orbital angular momentum $\tilde{l} = l - 1$ and pseudo spin $\tilde{s} = s = \frac{1}{2}$, these degenerate states can be expressed conveniently in terms of the pseudospin doublet with $j = \tilde{l} \pm \tilde{s}$. This concept is originally found in spherical nuclei, but later proved to be a good approximation in deformed nuclei as well.[2] Recently, many similar degeneracies were also observed among rotational bands, such as, ^{108}Tc,[3] ^{128}Pr,[4] and ^{189}Pt.[5] In this article, we report the experimental results on rotational band structures in ^{118}Sb. A pair of pesudospin partner bands have been observed in the present work.

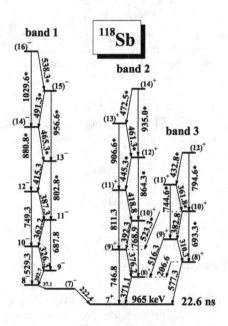

Fig. 1. Rotational band structures of ^{118}Sb proposed in the present work. New transitions observed in the present work are denoted with asterisks.

2. Experiment

High spin states of ^{118}Sb were populated in the ^{116}Cd(^{7}Li, 5n) fusion-evaporation reaction. The ^{7}Li beam was provided by the HI-13 tandem accelerator at the China Institute of Atomic Energy in Beijing. The beam energy of 50 MeV was chosen. The ^{116}Cd target was a self-supporting foil of 2.5 mg/cm^2 in thickness. Two fold $\gamma - \gamma$ coincident events were collected using an array of twelve Compton-suppressed HPGe detectors and two planar HPGe detectors. Approximately 2×10^8 γ-γ coincidence events were collected during this experiment. The data were sorted offline into symmetry matrix and DCO matrix. This data set had been utilized previously to establish the level scheme of ^{118}Sn.[6]

3. Result and discussion

Partial level scheme of ^{118}Sb obtained from the present work is shown in Fig. 1. As shown in Fig. 1, three strongly coupled bands are constructed and labeled as 1, 2, and 3. Band 1 is the most strongly populated of the rotational bands, and has been already assigned to the $\pi g_{9/2}^{-1} \otimes \nu h_{11/2}$ configuration.[7] The second band (band 2) built on the 22 ns 7^+ isomeric state,[8] and was proposed to be built predominantly on the $\pi g_{9/2}^{-1} \otimes \nu d_{5/2}$ configuration based on the magnetic moment measurements.[9] In the present work, a new coupled band structure (labeled 3) are identified, and four linking transitions between the bands 2 and 3 have been also observed.

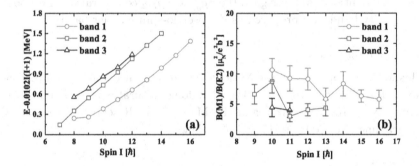

Fig. 2.   (color online) Excitation energy of the states relative to a rigid rotor (a) and $B(M1)/B(E2)$ ratios (b) for bands 1, 2, and 3 as a function of spin.

The excitation energies of the states in ^{118}Sb are plotted in Fig. 2(a) versus spin relative to a rigid-rotor reference. As shown in Fig. 2(a), the levels with the same spin and parity in the two bands 2 and 3 lie very close in energy. Furthermore, the experimental $B(M1)/B(E2)$ ratios for bands 1−3 were also deduced from the γ−intensities, and shown in Fig. 2(b). The intraband $B(M1)/B(E2)$ ratios for bands 2 and 3 are close to each other at the whole spin range. Thus, bands 2 and 3 may be expected as the chiral doublet bands, which had already been reported in the $A \sim 130$[10] mass region. In order to confirm or reject the hypothesis of chiral doublet bands, the configuration-fixed constrained triaxial relativistic mean-field (RMF) approaches[11] were applied to determine the quadrupole deformations. An axial prolate deformation ($\beta_2 \sim 0.23$, $\gamma \sim 0°$) are obtained from the present RMF approaches corresponding to the $\pi g_{9/2}^{-1} \otimes \nu d_{5/2}$ configuration. Therefore, the chiral speculation can be ruled out because the calculated shape turns out to be axial for the band 2. The energy difference between the band heads of band 2 ($I = 7\ \hbar$) and band 3 ($I = 8\ \hbar$) is 577 keV, which is close to the energy difference between the Nilsson states $g_{7/2}[402]5/2^+$ and $d_{5/2}[422]3/2^+$ in this mass region. Hence, we suggest that band 3 built on the $\pi g_{9/2}^{-1} \otimes \nu g_{7/2}$ configuration. Based on the present configuration assignments, bands 2 and 3 should be a pair of pseudospin partner bands.

4. Summary

High-spin states of ^{118}Sb was investigated using the ^{116}Cd(^{7}Li,5n) fusion-evaporation reaction at a beam energy of 50 MeV. The previously known band structures have been considerably extended. One new rotational band has been identified, and assigned the $\pi g_{9/2}^{-1} \otimes \nu g_{7/2}$ configuration. Two positive parity bands are proposed as the pseudo-spin partner bands.

Acknowledgments

This work is supported by the National Natural Science Foundation (Grant Nos. 10875074 and 11005069), the Shandong Natural Science Foundation (Grant No. ZR2010AQ005), and the Major State Research Development Programme (No. 2007CB815005) of China.

References

1. A. Arima, M. Harvey, and K. Shimizu, *Phys. Lett. B* **30**, 517 (1969).
2. J. Meng, K. Sugawara-Tanabe, S. Yamaji, P. Ring, and A. Arima, *Phys. Rev. C* **58**, R628 (1998).
3. Q. Xu *et al.*, *Phys. Rev. C* **78**, 064301 (2008).
4. C. M. Petrache *et al.*, *Phys. Rev. C* **65**, 054324 (2002).
5. W. Hua *et al.*, *Phys. Rev. C* **80**, 034303 (2009).
6. S. Y. Wang *et al.*, *Phys. Rev. C* **81**, 017301 (2010).
7. S. Vajda *et al.*, *Phys. Rev. C* **27**, 2995 (1983).
8. M. Fayez-Hassan *et al.*, *Nucl. Phys* **A624**, 401 (1997).
9. M. Ionescu-Bujor, A. Iordanescu, G. Pascovici and V. Sabaiduc, *Phys. Lett. B* **200**, 259 (1988).
10. S. Y. Wang, Y. Z. Liu, T. Komatsubara, Y. J. Ma, and Y. H. Zhang, *Phys. Rev. C* **74**, 017302 (2006), and references therein.
11. J. Meng, J. Peng, S. Q. Zhang and S. G. Zhou, *Phys. Rev. C* **73**, 037303 (2006).

STUDY OF ELASTIC RESONANCE SCATTERING AT CIAE[*]

Y.B. WANG[†], S.J. JIN, B.X. WANG, X. LIU, X.X. BAI,
Z.H. LI, G. LIAN, B. GUO, S. ZENG, J. SU, S.Q. YAN, X. QIN, Y.J.
LI, E.T. LI, and W.P. LIU

*Department of Nuclear Physics, China Institute of Atomic Energy(CIAE), P.O. BOX
275(10), Beijing102413, P.R. China*
[†]Email: ybwang@ciae.ac.cn

The proton elastic resonance scattering induced by radioactive secondary beams has been studied in inverse kinematics at CIAE since 2005. The so-called thick-target method is applied to obtain the excitation function in a one-shot experiment of a single beam energy. Up to now, several light nuclei systems including ^{13}N+p, ^{17}F+p, and ^{6}He+p have been successfully investigated with this method. A general introduction of the method and a summary of the results are presented in this paper.

Keywords: Nuclear Astrophysics; radioactive secondary beam; elastic resonance scattering; thick-target method.

1. Introduction

Nuclear astrophysics deals with the mechanism of the energy generation and element synthesis in various cosmic environments, by combining the astronomic observations with laboratory nuclear physics studies [1,2]. Therefore, understanding the underlying physical processes of star evolution relies on both, and in particular, studies of the relevant reaction network are of prime importance in reproducing the observed element abundances. The nuclear inputs need, according to different astronomic processes, mass, half-life, decay and reaction rates, separation energy etc. Among these, the single-particle or α-cluster resonances close to the threshold are of fundamental importance in nuclear astrophysics, since their dominant role in determining the reaction rates and therefore the thermonuclear runway [3,4].

The properties of the resonances are normally obtained by analyzing an experimental excitation function. For a stable projectile and target combination,

[*]This work is supported by the Major State Basic Research Development Program under Grant No. 2007CB815003, the National Natural Science Foundation of China under Grant Nos. 11021504, 10875173, 10735100.

the excitation function can be measured precisely by changing the beam energy in small steps to hit a very thin target. For nuclei with short lifetimes, the so-called thick-target technique with inverse kinematics is a novel way for the experimental excitation function [5-9].

The proton resonance scattering induced by radioactive secondary beams has been investigated [10] with the thick-target method at CIAE since 2005. The method has an advantage that allows one to obtain the excitation function in a one-shot experiment of a single beam energy. Several light nuclei systems including ^{13}N+p, ^{17}F+p, and ^{6}He+p have been successfully investigated, and a summary of the experimental results is given in this paper.

2. Experiment

The experiments were carried out at the radioactive secondary beam line [11] of the HI-13 Tandem accelerator laboratory, Beijing. In order to fulfill the requirements of the thick-target experiments, certain measures were taken to improve the secondary beam intensity in particular. As an example, the intensity of ^{13}N was enhanced by nearly an order of magnitude from about 1200 particles/s to 10000 particles/s. The secondary beam production conditions are listed in Table 1.

Table 1. The secondary beam production conditions.

Secondary beam	Production reaction	Energy (MeV)	Purity (%)	Intensity (max. pps)
^{6}He	^{2}H(^{7}Li,^{6}He)^{3}He	36.6	90	2000
^{13}N	^{2}H(^{12}C,^{13}N)n	47.8	75	10000
^{17}F	^{2}H(^{16}O,^{17}F)n	55.5	75	6000

Similar experimental setup was used in the series of thick-target measurements, and a schematic layout is shown in Figure 1. The secondary beam was collimated by a ϕ9-ϕ5 mm collimator complex to limit the beam spot size. A 13.2 µm thin ORTEC silicon detector was used to monitor the secondary beam in front of the target. $(CH_2)_n$ foils with thickness enough to stop the secondary beam ions were used as the reaction target, while carbon foils with equivalent thickness served for the background evaluation. The recoil protons were detected by a ΔE-E telescope, which is composed of a 60-70µm double-sided silicon strip detector(DSSSD) and a 1mm multi-guard silicon quadrant(MSQ). The telescope was placed at 15° instead of 0° to avoid the direct bombardment of the DSSSD by leaked light-ion components.

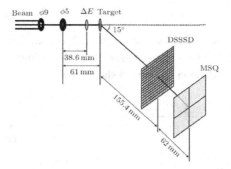

Fig. 1. Schematic layout of the experimental setup.

3. Results

3.1. $^{12}C+p$

In order to check the feasibility of the thick-target experiment at the local radioactive secondary beam facility and the performance of the detector system, the ^{12}C+p elastic resonance scattering was measured with conditions similar to those of secondary beam [12]. The ^{12}C beam was scattered into the secondary beam line, with the setup shown in Figure 1, the excitation function of ^{12}C+p elastic resonance scattering was obtained as shown in Figure 2. One can see that the theoretical curve agrees well with the experimental excitation function, indicating that the experimental excitation function can be well reproduced with the known resonance parameters.

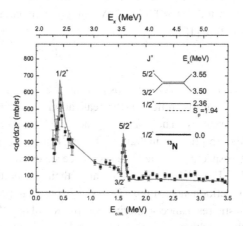

Fig. 2. Experimental excitation function for the ^{12}C+p elastic resonance scattering, the *R*-matrix fitting calculations with compiled resonance parameters is shown as the solid line.

3.2. $^{13}N+p$

The ^{13}N+p elastic resonance scattering was then measured for the missing 0^- state in ^{14}O and for the resonance parameters of the low-lying ^{14}O resonances. The experimental excitation function is shown in Figure 3.

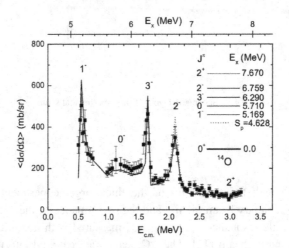

Fig. 3. The experimental excitation function for ^{13}N+p scattering. The inset shows the observed ^{14}O levels in the experiment. The best fitting result is indicated by the solid line.

The experimental excitation function covers the energy interval of $E_{c.m.} \approx$ 0.5-3.2 MeV, in which five low-lying proton resonance states in ^{14}O are observed [13,14]. The 0^- state shows up at 5.71 MeV in ^{14}O as a broad s-wave resonance, with a width of 400(45) keV, in good agreement with a work published shortly before [15].

3.3. $^{17}F+p$

Understanding the most violent explosions in the universe, recognized as supernovae or X-ray bursts, relies on the realization of the ignition conditions and thermonuclear runway. The commonly accepted breakout paths from hot CNO cycle to the rp-process include ^{14}O(α,p)^{17}F(p,γ)^{18}Ne and ^{15}O(α,γ)^{19}Ne reactions [16,17], respectively. The ^{18}Ne resonance states are relevant to both the ^{14}O(α,p)^{17}F and ^{17}F(p,γ)^{18}Ne reactions, and their properties will directly determine the competition against the ^{15}O(α,γ)^{19}Ne breakout path.

The ^{17}F+p elastic resonance scattering was measured at a ^{17}F beam energy of 55.5 MeV, and the resulting excitation function is shown in Figure 4. Only two

s-wave resonance states in [18]Ne were observed [18,19], with deduced resonance parameters in consistence with previously reported ones [20].

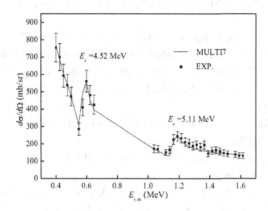

Fig. 4. The experimental excitation function for [17]F+p scattering.

3.4. [6]He+p

Study of [6]He+p elastic resonance scattering is motivated by nuclear structural issues, in particular, to search for the isospin analog state of the first excited state of the neutron drip-line nucleus [7]He. The measurement was done with a [6]He beam of the energy of 36.6 MeV, and the experimental data are being analyzed currently.

4. Summary

The thick-target method has been applied for studies of proton elastic resonance scattering induced by radioactive secondary beams. Experimental excitation functions for several light nuclei system have been measured, and the resonance parameters are deduced via R-matrix analysis. The preferable population of low-lying s-wave resonances in the compound nucleus is obvious in the studied system, and these resonance states are of close relevance to the proton capture reactions in different hydrogen burning processes at various cosmic environments. Therefore, the deduced resonance parameters are useful for elucidating the properties of resonance levels of astrophysical interests.

References

1. E. M. Burbidge, G. R. Burbidge, W. A. Fowler *et al.*, Rev. Mod. Phys. **29**, 547(1957).

2. C. Iliadis, *Nuclear Physics of Stars*, (Wiley, Weinheim, 2007).
3. R. K. Wallace and S. E. Woosley, Astrophys. J. Suppl. **45**, 389(1981).
4. L. Buchmann, J. M. D'Auria and P. McCorquodale, Astrophys. J. **324**, 953(1988).
5. L. Axelsson, J. G. Borgem, S. Fayans *et al.*, Phys. Rev. **C54**, R1511(1996).
6. T. Teranishi, S. Kubono, S. Shimoura *et al.*, Phys. Lett. **B556**, 27(2003).
7. C. Ruiz, T. Davinson, F. Sarazin *et al.*, Phys. Rev. **C71**, 025802(2005).
8. H. Yamaguchi, Y. Wakabayashi, S. Kubono *et al.*, Phys. Lett. **B672**, 230(2009).
9. J. P. Mitchell, G. V. Rogachev, E. D. Johnson *et al.*, Phys. Rev. **C82**, 011601R(2010).
10. Y. B. Wang, B. X. Wang, X. X. Bai *et al.*, HEP& NP **30**(Suppl. II), 202(2006) (in Chinese).
11. X. X. Bai, W. P. Liu, J. J. Qin *et al.*, Nucl. Phys. **A588**, 273c(1995).
12. X. Qin, Y. B. Wang, X. X. Bai *et al.*, Chin. Phys. **C32**, 957(2008).
13. Y. B. Wang, B. X. Wang, X. Qin *et al.*, Phys. Rev. **C77**, 044304(2008).
14. Y. B. Wang, X. Qin, B. X. Wang *et al.*, Chin. Phys. **C33**, 181(2009).
15. T. Teranishi, S. Kubono, H. Yamaguchi *et al.*, Phys. Lett. **B650**, 129(2007).
16. M. Wiescher, J. Görres, H. Schatz, J. Phys. **G25**, R133(1999).
17. H. Schatz, A. Aprahamian, J. Görres *et al.*, Phys. Rep. **294**, 167(1998).
18. J. S. Jin, Y. B. Wang, B. X. Wang *et al.*, Chin. Phys. Lett. **27**, 032102(2010).
19. Y. B. Wang, B. X. Wang, J. S. Jin *et al.*, Nucl. Phys. **A834**, 100c(2010).
20. J. Gómez del Campo, A. Galindo-Uribarri, J. R. Beene *et al.*, Phys. Rev. Lett. **86**, 43(2001).

SYSTEMATIC STUDY OF SURVIVAL PROBABILITY OF EXCITED SUPERHEAVY NUCLEI

C. J. XIA and B. X. SUN

*College of Applied Sciences, Beijing University of Technology,
Beijing, 100124, China*

E. G. ZHAO and S. G. ZHOU*

*Institute of Theoretical Physics, Chinese Academy of Sciences,
Beijing, 100190, China;
Center of Theoretical Nuclear Physics, National Laboratory of Heavy Ion Accelerator,
Lanzhou, 730000, China
E-mail: sgzhou@itp.ac.cn

In this contribution we present some results of a study of the stability of excited superheavy nuclei (SHN) with $100 \leq Z \leq 134$ against neutron emission and fission. We give a systematic investigation of the survival probability against fission in the 1n-channel of these SHN. The neutron separation energies and shell correction energies are consistently taken from the finite range droplet model which predicts an island of stability of superheavy nuclei around $Z = 115$ and $N = 179$. It is found that this island of stability persists for excited superheavy nuclei in the sense that the calculated survival probabilities in the 1n-channel of excited superheavy nuclei at the optimal excitation energy are maximized around $Z = 115$ and $N = 179$. This indicates that the survival probability in the 1n-channel is mainly determined by the nuclear shell effects.

Keywords: Survival probability; superheavy elements; island of stability; neutron emission; fission.

1. Introduction

The importance of quantum shell effects in stabilizing heavy nuclei was realized and the existence of superheavy elements (SHN) was predicted in 1960s.[1–3] Since then, many achievements have been made in exploring the island of stability of superheavy nuclei.[4–6]

Besides the stability of superheavy nuclei in their ground states, the stability of excited superheavy nuclei is also an important issue. On the one hand, it helps us in understanding the stability behavior of a superheavy nucleus against excitation. On the other hand, the survival probability of an excited compound nucleus against processes in which the charge number is changed is directly related with the stability of an excited superheavy nucleus in various channels. The survival probability is an important factor in the the study of the synthesis mechanism of superheavy elements. Most of the calculations are focused on the stability of superheavy compound nuclei formed in cold or hot fusion reactions.[7,8]

In order to study the influence of the shell effects on the stability of superheavy nuclei with excitation, we have carried out a systematic investigation of the stability of excited superheavy nuclei.[9] The stability of excited superheavy nuclei with $100 \leq Z \leq 134$ against neutron emission and fission was studied by using a statistical model. As an example, a systematic study of the survival probability against fission in the 1n-channel of these superheavy nuclei was made. In this contribution, we briefly present some of these results.

2. Formalism

An excited superheavy compound nucleus can decay via fission or emitting photon(s), neutron(s), proton(s) or light-charged particle(s) like α-particle. Among all these channels, the most favorable ones are fission and neutron(s) emission. In the present study, we mainly focus on these two channels.

The survival probability of an excited superheavy compound nucleus with excitation energy E^* and spin J in the 1n-channel is calculated by[7–10]

$$W_{\mathrm{sur}}(E^*, J) = P_{1\mathrm{n}}(E^*, J)\frac{\Gamma_{\mathrm{n}}(E^*, J)}{\Gamma_{\mathrm{n}}(E^*, J) + \Gamma_{\mathrm{f}}(E^*, J)}, \tag{1}$$

where $P_{1\mathrm{n}}$ is the realization probability for one neutron emission, Γ_{n} is the width for the neutron emission and Γ_{f} is the fission width. Further details of the calculation can be found in Ref.[9]

3. Results and discussions

There are many different predictions for the ground state and saddle point properties of the superheavy nuclei. Since both the neutron emission and fission processes are connected closely to the shell structure, in order to study the influence of the shell effects on the stability of superheavy nuclei with excitation, one should take the nuclear property parameters for calculating the neutron emission and fission consistently from one single model. In the present work, the properties of superheavy nuclei with $100 \leq Z \leq 134$ are taken from predictions of the finite range droplet model.[11]

The survival probability of an excited superheavy nucleus with $J = 0$ in the 1n-channel calculated from Eq. (1), as a function of the excitation energy E^*, shows an anti-parabolic shape. We take the maximal value in the $W_{sur} \sim E^*$ curve for each superheavy nucleus and define the corresponding E^* as the optimal excitation energy.

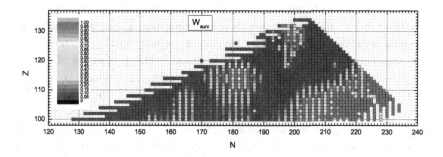

Fig. 1. The survival probability against fission of excited superheavy nuclei in the 1n-channel with $100 \leq Z \leq 134$ at the optimal excitation energy and $J = 0$.[9]

In Fig. 1 the survival probability of excited superheavy nuclei with $100 \leq Z \leq 134$ at the optimal excitation energies and $J = 0$ are given. There are three islands with larger survival probability in Fig. 1 which roughly correspond to the islands of superheavy nuclei with higher fission barriers, namely the island of stability centers around $Z = 115$ and $N = 179$ due to the quantum shell effects, the island at $Z = 108$ and $N = 162 \sim 164$ due to the deformed sub-shells, and the mass region around $Z = 130$ and $N = 198$. This indicates that for a superheavy nucleus in these three mass regions, the fission width is quite small due to the high fission barrier. In the very neutron-rich region, the neutron separation energy is small which results in a large neutron emission width. Therefore the survival probability in the 1n-channel becomes larger in the very neutron-rich region if one only includes the fission and neutron emission processes. For these nuclei, other decay channels should be taken into account. The survival probability of superheavy nuclei in the 1n-channel shows clear odd-even effects which is mainly from the odd-even effects in the neutron separation energy.

From the above discussions, one can conclude that the shell effects also plays important roles in the stability behavior and in the survival probability of excited superheavy nuclei. Besides the neutron separation energy and the fission barrier which are directly related to the shell effects, there are some other parameters which are also very important in the calculation of the survival probability. For example, the parameters for the level density influence the decay width very much.[8]

4. Summary

In order to study the influence of the shell effects on the stability of superheavy nuclei with excitation, the stability of excited superheavy nuclei with $100 \leq Z \leq 134$ against neutron emission and fission is studied and a systematic investigation of the survival probability in the 1n-channel of these superheavy nuclei with $J = 0$ is made.[9] The properties of superheavy nuclei including the neutron separation energies and shell correction energies are consistently taken from predictions of the finite range droplet model.[11] It is found that the islands of stability of superheavy nuclei in their ground state, e.g., the one around $Z = 115$ and $N = 179$, persist for excited superheavy nuclei in the sense that the calculated survival probabilities in the 1n-channel of excited superheavy nuclei on these islands are maximized. This indicates that the survival probability is mainly determined by the nuclear shell effects. In addition, the survival probability of superheavy nuclei in the 1n-channel shows clear odd-even effects.

Acknowledgments

We thank G. G. Adamian, N. V. Antonenko, A. Nasirov and A. S. Zubov for fruitful discussions and suggestions. This work was partly supported by NSFC (10705014, 10775012, 10875157, 10975100, and 10979066), MOST (973 project 2007CB815000), CAS (KJCX2-EW-N01 and KJCX2-YW-N32), and Supercomputing Center, CNIC, CAS.

References

1. W. D. Myers and W. J. Swiatecki, *Nucl. Phys.* **81**, 1 (1966).
2. A. Sobiczewski, F. Gareev and B. Kalinkin, *Phys. Lett.* **22**, 500 (1966).
3. H. Meldner, *Arkiv Fysik* **36**, 593 (1967).
4. S. Hofmann and G. Münzenberg, *Rev. Mod. Phys.* **72**, 733 (2000).
5. Y. Oganessian, *J. Phys. G: Nucl. Phys.* **34**, R165 (2007).
6. K. Morita, K. Morimoto, D. Kaji, T. Akiyama, S.-i. Goto, H. Haba, E. Ideguchi, R. Kanungo, K. Katori, H. Koura, H. Kudo, T. Ohnishi, A. Ozawa, T. Suda, K. Sueki, H.-S. Xu, T. Yamaguchi, A. Yoneda, A. Yoshida and Y.-L. Zhao, *J. Phys. Soc. Jpn.* **73**, 2593 (2004).
7. G. G. Adamian, N. V. Antonenko, S. P. Ivanova and W. Scheid, *Phys. Rev. C* **62**, 064303 (2000).
8. A. Zubov, G. Adamian and N. Antonenko, *Phys. Part. Nucl.* **40**, 847 (2009).

9. C.-J. Xia, B.-X. Sun, E.-G. Zhao and S.-G. Zhou, arXiv: 1101.2725 [nucl-th], to be published in *Science China - Physics, Mechanics and Astronomy*.
10. Z.-Q. Feng, G.-M. Jin, F. Fu and J.-Q. Li, *Nucl. Phys. A* **771**, 50 (2006).
11. P. Möller, J. R. Nix, W. D. Myers and W. J. Swiatecki, *At. Data Nucl. Data Tables* **59**, 185 (1995).

ANGULAR MOMENTUM PROJECTION OF THE NILSSON MEAN-FIELD PLUS NEAREST-ORBIT PAIRING INTERACTION MODEL

MING-XIA XIE[†], QI TAN, LU JIA, and FENG PAN

Department of Physics, Liaoning Normal University, Dalian, 116029, China
[†]*E-mail: xiemingxia@sina.com*

J. P. DRAAYER

*Department of Physics and Astronomy, Louisiana State University,
Baton Rouge, LA 70803-4001, USA*

A projected Hamiltonian with definite angular momentum from the exactly solvable deformed mean-field plus nearest-orbit pairing model is proposed to study whether the deformed mean-field plus nearest-orbit pairing model can be used to describe low-lying spectra of nuclei reasonably.

Keywords: Nilsson mean-field; nearest-orbit pairing interaction; angular momentum projection.

1. Introduction

The mean field plus pairing model was previously studied by using the BCS approximation.[1] It is well known that exact solution to the problem can be obtained by using the Gaudin-Richardson method. However, with increasing the number of orbits and valence nucleon pairs, the solutions turn out to be much more complicated. To avoid such complication, the nearest-orbit pairing interaction was considered, which has been proven to be a good approximation for the pairing interaction in deformed nuclei.[1] However, it is still not possible to describe low-lying states in the model because angular momentum is not a conserved quantity due to the intrinsic deformation. Hence, angular momentum projection for the model is necessary in order to project the system into the physical subspace with definite angular momentum. In addition, particle number nonconservation effects occurring in the BCS approximation treatment is avoided in the angular momentum projected Nilsson mean-field plus nearest-orbit pairing model because the pairing interaction is treated exactly.

2. Angular momentum projection of the model

The Hamiltonian of the mean-field plus nearest-orbit pairing model for either proton or neutron part can be written as:[2]

$$\hat{H} = \sum_i \varepsilon_i^\rho + \sum_{<i,j>}' t_{ij}^\rho b_i^\dagger(\rho) b_j(\rho), \tag{1}$$

where the first term only involves single-particle energies of unpaired particles, and the prime in the summation of the second term indicates that the sum is restricted to the orbits not occupied by those unpaired particles, $\rho = \pi$ or ν for the proton or neutron part, $b_i^\dagger(\rho)$ $(b_i(\rho))$ are pair creation (annihilation) operators, $t_{\alpha\beta}^\rho = 2\epsilon_\alpha^\rho \delta_{\alpha\beta} + A e^{-B(\varepsilon_\alpha - \varepsilon_\beta)^2}$, in which ε_α and ε_β are single-particle energies of orbit α and orbit β respectively taken from the Nilsson model, $A < 0$ and $B > 0$ are real parameters, and the sum $< i, j >$ only runs over the same and the nearest orbits. As shown in Ref. [2], the μ-pair eigenstates of Eq.(1) can be written as

$$|\mu(\rho); \xi(\rho); n_f^\rho\rangle = \sum_{i_1 < i_2 \cdots < i_{\mu(\rho)}} C_{i_1 i_2 \cdots i_{\mu(\rho)}}^{\xi(\rho)} b_{i_1}^\dagger(\rho) b_{i_2}^\dagger(\rho) \cdots b_{i_{\mu(\rho)}}^\dagger(\rho) |n_f^\rho\rangle, \tag{2}$$

where $|n_f^\rho\rangle$ is the pairing vacuum state satisfying

$$b_i(\rho)|n_f^\rho\rangle = 0. \tag{3}$$

In order to project the Hamiltonian of Eq. (1) into a physical subspace with good angular momentum, the projected Hamiltonian with given angular momentum J and its third projection in the intrinsic frame K can be written as

$$\tilde{H}_K^J = \hat{P}_K^J \hat{H} \hat{P}_K^J, \tag{4}$$

where

$$\hat{P}_K^J = \sum_\tau |\tau JK\rangle\langle\tau JK|, \tag{5}$$

in which $\{|\tau JK\rangle\}$ can be expanded in terms of eigenstates $\{|k\xi\rangle\}$ of the deformed mean-field plus pairing model Hamiltonian of Eq. (1), where τ is an additional quantum number used to label different states with the same J, and $\{|k\xi\rangle\}$ is a complete set of normalized orthogonal eigenstates obtained from the mean-field plus pairing model, in terms of which $|\tau JK\rangle$ can be expanded as

$$|\tau JK\rangle = \sum_\xi \langle k\xi|\tau JK\rangle|k\xi\rangle = \sum_\xi g_\xi^{\tau JK}|k\xi\rangle, \tag{6}$$

where k is total valence particle number in the corresponding system, $g_\xi^{\tau JK}$ is the expansion coefficient which is the key to determine $\{|\tau JK\rangle\}$. In order to determine the expansion coefficient $g_\xi^{\tau JK}$, we adopt the usual angular momentum projection operator:

$$P_K^J = \frac{2J+1}{8\pi^2} \int d\alpha \sin\beta d\beta d\gamma D_{KK}^J(\alpha, \beta, \gamma) e^{i\alpha \hat{J}_z} e^{i\beta \hat{J}_y} e^{i\gamma \hat{J}_z}, \tag{7}$$

which satisfies

$$P_K^J|\tau JK\rangle = |\tau JK\rangle. \tag{8}$$

Combining Eq. (6) and Eq. (8), we obtain the following linear equation for the coefficients $g_\xi^{\tau JK}$:

$$g_{\xi'}^{\tau JK} = \sum_\xi g_\xi^{\tau JK} \langle k\xi'|P_K^J|k\xi\rangle. \tag{9}$$

Then, the eigenstates $|\eta JK\rangle$ of $\tilde{H}_K^J$ can be expressed as

$$|\eta JK\rangle = \sum_\tau a_\tau^\eta|\tau JK\rangle, \tag{10}$$

where a_τ^η is the corresponding expansion coefficient determined by the following eigen-equation:

$$\tilde{H}_K^J|\eta JK\rangle = E_{JK}^\eta|\eta JK\rangle. \tag{11}$$

Since the quantum number $\eta \equiv \eta_K$ is K dependent, the final eigenstates in the laboratory frame $|\eta JK\rangle_L$ should be expressed in terms of those in the intrinsic frame $|\eta JK\rangle$ as

$$|\eta_K JM\rangle_L \sim D_{MK}^{J\,*}(R)|\eta_K JK\rangle \tag{12}$$

with the corresponding eigenenergies the same as those given by Eq. (11). Contrary to the triaxial rotor, however, there is no D_2 symmetry restriction for Eq. (12) except the time reversal symmetry with $K \rightleftarrows -K$. Therefore, both even and odd K values are allowed in Eq. (12).

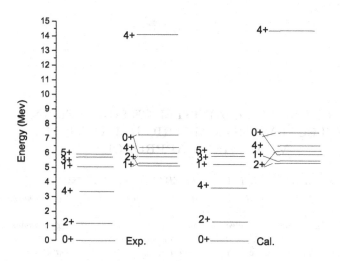

Fig. 1. Calculated Low-lying levels of ^{22}Mg and comparison with the experimental data.[3]

3. Application to ^{22}Mg

As a simple example to test the theory, low-lying energy levels of ^{22}Mg are calculated according to Eq. (12). As shown in Fig. 1, the low-lying levels of ^{22}Mg are well reproduced by the angular momentum projected Nilsson mean-field plus nearest orbit pairing model.

Acknowledgments

Support from the U.S. National Science Foundation (PHY-0500291 & OCI-0904874), the Southeastern Universities Research Association, the Natural Science Foundation of China (10775064), the Liaoning Education Department Fund (2007R28), the Doctoral Program Foundation of State Education Ministry of China (20102136110002), and the LSU–LNNU joint research program (9961) is acknowledged.

References

1. P. Ring and P. Schuck, *The Nuclear Many-Body Problem* (Springer-Verlag, Berlin, 1980).
2. Feng Pan and J. P. Draayer, *J. Phys. A* **33** (2000) 9095.
3. LBNL-Lund collaboration, Isotope Explorer version 3.0, ie.lbl.gov/ensdf/.

POSSIBLE SHAPE COEXISTENCE FOR ^{152}Sm IN A REFLECTION-ASYMMETRIC RELATIVISTIC MEAN-FIELD APPROACH

W. ZHANG[1,2,*] Z. P. LI[3], S. Q. ZHANG[2] and J. MENG[4,2]

^{1}School of Electrical Engineering and Automation, Henan Polytechnic University,
Jiaozuo 454003, PRC
^{2}State Key Laboratory of Nuclear Physics and Technology, School of Physics,
Peking University, Beijing 100871, PRC
^{3}School of Physical Science and Technology, Southwest University,
Chongqing 400715, PRC
^{4}School of Physics and Nuclear Energy Engineering, Beihang University,
Beijing 100191, PRC
*E-mail: zw76@pku.org.cn

The potential energy surfaces of ^{152}Sm and its neighboring isotopes are investigated in a constrained reflection-asymmetric relativistic mean-field approach with parameter set PK1. According to the calculations, ^{152}Sm has a quadrupole-deformed minimum and an octupole-deformed minimum with similar energies. The low-lying 0^+ states in the spectrum can be understood as the manifestation of shape coexistence in (β_2, β_3) plane. Furthermore, the neutron gap with $N = 88$ and proton gaps with $Z = 62$ in the single-particle levels may be important for this coexistence.

Keywords: Shape coexistence; octupole degree of freedom; relativistic mean-field.

Nuclear shape is governed by a delicate interplay of the macroscopic liquid-drop properties of nuclear matter and microscopic shell effects and is therefore regarded as a very sensitive probe of the underlying nuclear structure. An atomic nucleus with different configurations can behave as different shapes. Previous studies already show that nuclei in some mass regions indeed have the phenomena of shape coexistence. In these nuclei, there are at least two clearly distinguishable shapes, e.g., spherical versus deformed or prolate versus oblate. In addition, the energies of these shapes are nearly degenerate and the mixings of their configurations are very weak. Nuclei around the neutron-deficient Kr and Sr have been considered as the most favorable candidates for the presence of shape coexistence. For an even-even nucleus, the presence of low-lying 0^+ states in the spectrum is usually interpreted as the manifestation of shape coexistence.[1] Various theoretical models using Nilsson-Strutinsky approach,[2] improved microscopic-macroscopic model,[3] self-consistent non-relativistic mean-field models[4,5] as well as relativistic models[6] confirm the presence of oblate and prolate minima in the deformation energy surfaces of some light Kr, Sr, and Zr isotopes. Particularly for ^{82}Sr, the experimental bands are interpreted to have prolate, oblate, or triaxial shapes by the Woods-Saxon cranking calculations with pairing.[7] The single-particle levels of the underlying mean field, especially near the nucleon number $N = 40$, may play important roles for such phenomenon.[8]

For heavier nuclei, it is shown that ^{152}Sm and other $N = 90$ isotones are the empirical examples of the analytic description of nuclei at the critical-point of the first-order shape/phase transition between spherical $U(5)$ and axially deformed $SU(3)$ shapes.[9–11] More references can be found in Ref.[12,13].

Interestingly, ^{152}Sm lies around one of the regions with strong octupole correlations. Such regions correspond to either the proton or neutron numbers close to 34 ($1g_{9/2} \leftrightarrow 2p_{3/2}$ coupling), 56 ($1h_{11/2} \leftrightarrow 2d_{5/2}$ coupling), 88 ($1i_{13/2} \leftrightarrow 2f_{7/2}$ coupling), and 134 ($1j_{15/2} \leftrightarrow 2g_{9/2}$ coupling).[14] A variety of approaches have been applied to investigate the role of octupole degree of freedom in Sm and neighboring nuclear region.[15–17]

Recently, a new $K^\pi = 0^-$ octupole excitation band for ^{152}Sm is observed and a pattern of repeating excitations built on the 0_2^+ level similar to those built on the ground state emerges.[18] It is suggested that ^{152}Sm, rather than a critical-point nucleus, is a complex example of shape coexistence.[18]

Based on the bands, it is timely and necessary to investigate the Sm isotopes in a microscopic and self-consistent approach with the octupole degree of freedom. The newly-developed Reflection-ASymmetric Relativistic Mean-Field (RAS-RMF) approach is a good candidate for this purpose,[19] considering the remarkable success of RMF theory[20–22] in describing many nuclear phenomena related to stable nuclei, exotic nuclei as well as supernova and neutron stars. Recently, the RAS-RMF approach is applied to the well-known octupole deformed nucleus ^{226}Ra,[19] La isotopes,[23] Sm isotopes,[13] Ba isotopes,[24] and Th isotopes.[25] By investigating the potential energy surfaces of even-even $^{146-156}$Sm isotopes, it is suggested that the critical-point candidate nucleus ^{152}Sm marks the shape/phase transition not only from $U(5)$ to $SU(3)$ symmetry, but also from the octupole deformed ground state in ^{150}Sm to quadrupole deformed ground state in ^{154}Sm.[13]

In this proceeding, we will investigate the possible shape coexistence of quadrupole-deformed shape and octupole-deformed shape in ^{152}Sm. The single-particle levels will also be discussed.

The framework of the RAS-RMF approach can be found in Ref. [13] and references therein.

The properties of ^{152}Sm and its neighboring isotopes are calculated in the constrained RAS-RMF approach with parameter set PK1.[26] The parameter set PK1 is one of the best parameter sets available. The TCHO basis with 16 major shells for both fermions and bosons is used. The pairing correlation is treated by the BCS approximation with a constant pairing gap $\Delta = 11.2/\sqrt{A}$ MeV.

The binding energy, quadrupole and octupole deformations for ^{150,152,154}Sm are listed in Table 1. The binding energies are well reproduced within 0.2%. Moreover, a satisfied agreement is obtained for the quadrupole deformations.[13] For ^{152}Sm, two minima (denoted by ^{152}Sm$_{min1}$ and ^{152}Sm$_{min2}$) emerge: one octupole-deformed, the other quadrupole-deformed.

Table 1. The total binding energy (in MeV) as well as the quadrupole deformation β_2 and octupole deformation β_3 of the ground states of ^{150,152,154}Sm obtained in the constrained RAS-RMF approach with PK1, in comparison with the available experimental data.

Nucleus	E^{cal}	β_2^{cal}	β_3^{cal}	E^{exp}[27]	β_2^{exp}[28]
^{150}Sm	1241.59	0.18	0.13	1239.25	0.19
^{152}Sm$_{min1}$	1254.48	0.21	0.17	1253.10	0.31
^{152}Sm$_{min2}$	1254.13	0.29	0		
^{154}Sm	1267.76	0.32	0	1266.94	0.34

As an example, the contour plots of the binding energies in (β_2, β_3) plane for ^{150,152,154}Sm with parameter set PK1 are displayed in Fig. 1. Similar figures with parameter set NL3 can be found in Ref.[29].

The calculations show that for the ground states, ^{146,148}Sm are near spherical, ^{150}Sm octupole deformed and ^{154,156}Sm well deformed while ^{152}Sm marks the transition from octupole to quadrupole deformed.[13] For ^{152}Sm with PK1, a global minimum ^{152}Sm$_{min1}$ as well as a quadrupole minimum ^{152}Sm$_{min2}$ emerge. Note that the deformation $\beta_2 = 0.29$ for ^{152}Sm$_{min2}$ is quite close to the experimental value $\beta_2^{\mathrm{exp}} = 0.31$.[28] It is clear that the region near the quadrupole minimum is soft, e.g., for parameter set PK1, the

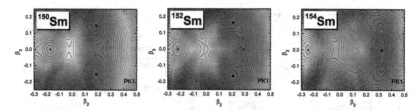

Fig. 1. The contour plots of total energies for ^{150,152,154}Sm in (β_2, β_3) plane obtained in RAS-RMF approach with parameter set PK1. The energy separation between contour lines is 0.25 MeV. The global minimum and other local minima are denoted by "■" and "•" respectively.

energy difference between two minima is 0.35 MeV with a 0.5 MeV barrier in between. After including the octupole degree of freedom, ^{152}Sm marks the shape/phase transition from the octupole deformed to quadrupole deformed case.

For ^{152}Sm, the pattern of repeating excitations discovered in Ref.[18] can be well understood from the PES obtained above. For the octupole minimum and the quadrupole minimum with PK1, if one performs the Generator Coordinate Method (GCM)[30,31] calculation with the PES, two low-lying states in the (β_2,β_3) plane with similar quadrupole deformation will be obtained, which are mixture of quadrupole and octupole deformation configurations. Based on these two states, the pattern of repeating excitations is expected. As mentioned in the introduction, such low-lying 0^+ states in the spectrum is usually interpreted as the manifestation of shape coexistence.[1] Based on our calculations, the quadrupole-deformed shape may coexist with the octupole-deformed shape.

With the octupole degree of freedom, a large energy gap with $N = 88$ near the Fermi surface for the neutron single-particle levels can be found while no obvious gaps near the Fermi surfaces can be found for protons.[13] To understand the possible shape coexistence, the single-particle levels in ^{152}Sm for the states between the minima ^{152}Sm$_{\text{min2}}$ and ^{152}Sm$_{\text{min1}}$ are shown in Fig. 2. The levels near the Fermi surface are labeled by Nilsson-like notations $\Omega[Nn_zm_l]$ of the largest component for state ^{152}Sm$_{\text{min2}}$. Note the single-particle levels are functions of β_3 here. For the state ^{152}Sm$_{\text{min2}}$ $(\beta_3 = 0)$, the neutrons and the protons near the Fermi surfaces partially occupy two corresponding adjacent levels with a relatively large energy gap in between, especially the proton gap with $Z = 62$. For the state ^{152}Sm$_{\text{min1}}$ $(\beta_3 \neq 0)$, the last two neutrons fully occupied the level above $N = 88$. Such gaps may be the clue for the shape coexistence.

238

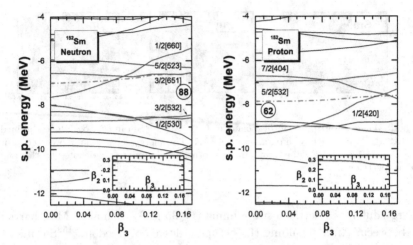

Fig. 2. (Color online) Single-particle levels of ^{152}Sm in RAS-RMF approach with PK1 as functions of β_3 for states between minima ^{152}Sm$_{min2}$ and ^{152}Sm$_{min1}$. The neutron (proton) single-particle levels are shown in left (right) panel. The dash-dot lines denote the corresponding Fermi surfaces. The levels near the Fermi surfaces are labeled by Nilsson-like notations $\Omega[Nn_z m_l]$ of the leading component for state ^{152}Sm$_{min2}$. The corresponding β_2 are shown in the inset.

In this proceeding, the potential energy surfaces of ^{152}Sm and its neighboring isotopes are obtained in (β_2, β_3) plane by the constrained RAS-RMF approach with parameter set PK1. Two minima exist for ^{152}Sm: one octupole-deformed with $(\beta_2, \beta_3)=(0.21,0.17)$, the other quadrupole-deformed with $(\beta_2, \beta_3)=(0.29, 0)$. The low-lying 0^+ states in the spectrum can be understood as the manifestation of shape coexistence in (β_2, β_3) plane. For the single-particle levels, the neutron gap of nearly 2 MeV with $N = 88$ for ^{152}Sm$_{min1}$ and proton gap of nearly 2 MeV with $Z = 62$ for ^{152}Sm$_{min1}$ may be important for the coexistence.

Acknowledgments

This work is supported in part by the National Basic Research Program of China under Grant No 2007CB815000, National Natural Science Foundation of China under Grant Nos 10775004, 10975007, and 10975008, China Postdoctoral Science Foundation, the Young Backbone Teacher Support Program of Henan Polytechnic University, the Funding for Henan Provincial Key Discipline: Detection Technology and Automation Equipment under Grant No 509923, and the Southwest University Initial Research Foundation Grant to Doctor (SWU110039).

References

1. J. L. Wood, K. Heyde, W. Nazarewicz *et al.*, *Phys. Rep.* **215**, 101 (1992).
2. F. Dickmann, V. Metag, and R. Repnow, *Phys. Lett. B* **38**, 207 (1972).
3. W. Nazarewicz, J. Dudek, R. Bengtsson *et al.*, *Nucl. Phys. A* **435**, 397 (1985).
4. P. Bonche, H. Flocard, P.-H. Heenen *et al.*, *Nucl. Phys. A* **443**, 39 (1985).
5. M. Girod, J. P. Delaroche, D. Gogny, and J. F. Berger, *Phys. Rev. Lett.* **62**, 2452 (1989).
6. G. A. Lalazissis and M. M. Sharma, *Nucl. Phys. A* **586**, 201 (1995).
7. C. Baktash *et al.*, *Phys. Lett. B*, **255**, 174 (1991).
8. S. J. Zheng, doctoral thesis: *Nuclear Deformations and Shape Coexistences* (Peking Univ., Beijing, 2008).
9. F. Iachello, N. V. Zamfir, and R. F. Casten, *Phys. Rev. Lett.* **81**, 1191 (1998).
10. F. Iachello, *Phys. Rev. Lett.* **87**, 052502 (2001).
11. R. F. Casten and N. V. Zamfir, *Phys. Rev. Lett.* **87**, 052503 (2001).
12. J. Meng, W. Zhang, S. G. Zhou, H. Toki, and L. S. Geng, *Eur. Phys. J. A* **25**, 23 (2005).
13. W. Zhang, Z. P. Li, S. Q. Zhang, and J. Meng, *Phys. Rev. C* **81**, 034302 (2010).
14. P. A. Butler and W. Nazarewicz, *Rev. Mod. Phys.* **68**, 349 (1996).
15. W. Nazarewicz and S. L. Tabor, *Phys. Rev. C* **45**, 2226 (1992).
16. M. Babilon *et al.*, *Phys. Rev. C* **72**, 064302 (2005).
17. N. Minkov, P. Yotov, S. Drenska *et al.*, *Phys. Rev. C* **73**, 044315 (2006).
18. P. E. Garrett *et al.*, *Phys. Rev. Lett.* **103**, 062501 (2009).
19. L. S. Geng, J. Meng, and H. Toki, *Chin. Phys. Lett.* **24**, 1865 (2007).
20. P. Ring, *Prog. Part. Nucl. Phys.* **37**, 193 (1996).
21. D. Vretenar, A. V. Afanasiev, G. A. Lalazissis, and P. Ring, *Phys. Rep.* **409**, 101 (2005).
22. J. Meng, H. Toki, S. G. Zhou, S. Q. Zhang, W. H. Long, and L. S. Geng, *Prog. Part. Nucl. Phys.* **57**, 470 (2006).
23. N. Wang and L. Guo, *Sci. China Ser. G* **52(10)**, 1574 (2009).
24. W. Zhang, Z. P. Li, and S. Q. Zhang, *Chin. Phys. C* **34(8)**, 1094 (2010).
25. J. Y. Guo, J. Peng, and X. Z. Fang, *Phys. Rev. C* **82**, 047301 (2010).
26. W. H. Long, J. Meng, N. V. Giai, and S. G. Zhou, *Phys. Rev. C* **69**, 034319 (2004).
27. G. Audi, A. H. Wapstra, and C. Thibault, *Nucl. Phys. A* **729**, 337 (2003).
28. S. Raman, C. W. Nestor Jr., P. Tikkanen, *At. Data Nucl. Data Tables* **78**, 1 (2001).
29. W. Zhang, Z. P. Li, S. Q. Zhang, and J. Meng, *J. Phys. Con. Ser.* (INPC2010).
30. P. Ring and P. Schuck, *The Nuclear Many-body Problem* (Springer-Verlag, New York, 1980).
31. J. M. Yao, J. Meng, P. Ring, and D. Vretenar, *Phys. Rev. C* **81**, 044311 (2010).

NUCLEAR PAIRING REDUCTION DUE TO ROTATION AND BLOCKING

ZHEN-HUA ZHANG (张振华)

Institute of Theoretical Physics, Chinese Academy of Sciences, Beijing 100190, China

Nuclear pairing gaps of normally deformed nuclei are investigated by using the particle-number conserving formalism for the cranked shell model, in which the blocking effects are treated exactly. Both rotational frequency ω-dependence and seniority ν-dependence of the pairing gap are investigated.

Keywords: pairing gap; moment of inertia; pairing correlation; particle-number conserving method.

1. Introduction

Since the seminal article by Bohr, Mottelson, and Pines,[1] significant effects of nuclear pairing were established in fundamental nuclear properties.[2] Soon afterwards, the Bardeen-Cooper-Schrieffer (BCS) theory for metallic super-conductivity and quasiparticle (qp) formalism were transplanted in nuclear structure literature to treat nuclear pairing correlation.[3–5] However, along with their great successes, both BCS and HFB approximations for nuclear pairing suffer from serious problems.[6,7] One of them is the non-conservation of the particle-number. Because the number of nucleons in a nucleus is not very large ($n \sim 10^2$), particularly the number of valence nucleons ($n \sim 10$) dominating the nuclear low-lying excited states is very limited, the relative particle-number fluctuation, $\delta n/n$, is not negligible. Indeed, it was found that in all self-consistent solutions to the cranked HFB equation a pairing collapsing occurs for angular momentum I greater than a critical value I_c.[8] However, no pairing phase transition was found in the number-projected HFB (NHFB) calculations,[9,10] which definitely shows that the occurrence of nuclear pairing collapsing originates from particle-number non-conservation. In this article, we use the particle-number conserving (PNC) method[6,11] to investigate the pairing reduction, in which the blocking effects are treated exactly.

2. Rotational dependence of the nuclear pairing gaps

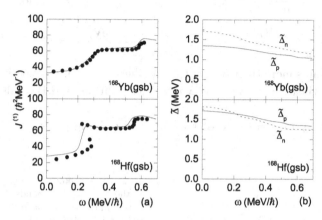

Fig. 1.   The MOIs and pairing gaps for the gsb's of ^{168}Yb and ^{168}Hf. (a) The experimental (calculated) MOIs are denoted by the solid circle (solid lines). The Nilsson parameters (κ, μ) and deformation $(\varepsilon_2, \varepsilon_4)$ are taken from.[12,13] The monopole and quadrupole pairing strengths (in MeV) for protons and neutrons are $G_n = 0.30$, $G_{2n} = 0.010$, $G_p = 0.29$ for ^{168}Yb; $G_n = 0.39$, $G_p = 0.35$ for ^{168}Hf. (b) The pairing gaps calculated by PNC method for protons (neutrons) are denoted by solid (dashed) lines.

The angular momentum dependence of pairing gaps $\tilde{\Delta}_n$ (neutrons) and $\tilde{\Delta}_p$ (protons) for the ground state band (gsb) of ^{168}Yb and ^{168}Hf have been calculated by the NHFB approach in Ref.[10] The pairing gap reductions in the observed angular momentum range for ^{168}Yb(gsb) and ^{168}Hf(gsb) calculated by NHFB are[9,10]

$$\frac{\tilde{\Delta}_n(I = 44\hbar)}{\tilde{\Delta}_n(I = 0)} \approx 48\%, \quad \frac{\tilde{\Delta}_p(I = 44\hbar)}{\tilde{\Delta}_p(I = 0)} \approx 63\% \quad \text{for } {}^{168}\text{Yb(gsb)},$$

$$\frac{\tilde{\Delta}_n(I = 34\hbar)}{\tilde{\Delta}_n(I = 0)} \approx 38\% \qquad\qquad \text{for } {}^{168}\text{Hf(gsb).} \quad (1)$$

For comparison, the ω-dependence of pairing gaps of ^{168}Yb(gsb) and ^{168}Hf(gsb) are calculated by the PNC formalism. The experimental MOIs $J^{(1)}$ for the gsb's are well reproduced by the PNC calculations. The calculated ω-dependence of the pairing gaps $\tilde{\Delta}_n$ and $\tilde{\Delta}_p$ are shown in Fig. 1(b). As expected, in both NHFB and PNC formalism no pairing phase transition from superfluidity to normal motion ($\tilde{\Delta} \to 0$) is found with increasing ω. In the observed rotational frequency range, the pairing gap reductions calculated in PNC formalism are

$$\frac{\tilde{\Delta}_n(\omega = 0.61\text{MeV}/\hbar)}{\tilde{\Delta}_n(\omega = 0)} \approx 70\%, \quad \frac{\tilde{\Delta}_p(\omega = 0.61\text{MeV}/\hbar)}{\tilde{\Delta}_p(\omega = 0)} \approx 80\% \text{ for } {}^{168}\text{Yb},$$

$$\frac{\tilde{\Delta}_n(\omega = 0.52\text{MeV}/\hbar)}{\tilde{\Delta}_n(\omega = 0)} \approx 70\%, \quad \frac{\tilde{\Delta}_p(\omega = 0.52\text{MeV}/\hbar)}{\tilde{\Delta}_p(\omega = 0)} \approx 83\% \text{ for } {}^{168}\text{Hf},(2)$$

which remains more than 70% of the bandhead value in all experimental ω range and the ω-dependence of $\tilde{\Delta}$ is weaker than the NHFB results.

3. Seniority dependence of the nuclear pairing gaps

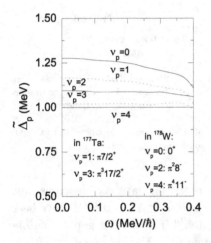

Fig. 2. The protons pairing gaps $\tilde{\Delta}_p$ for the $\nu_p = 1, 3$ bands in ^{177}Ta (dotted lines) and $\nu_p = 0, 2, 4$ configurations in ^{178}W (solid lines). ^{177}Ta: $\nu_p = 1$ band (gsb), $\pi 7/2^+[404]$; $\nu_p = 3$, $K^\pi = 17/2^+$ band at 1523 keV,[14] $\pi^3 17/2^+(7/2^+[404] \otimes 9/2^-[514] \otimes 1/2^-[541])$. ^{178}W: gsb, $\nu_p = 0$, $K^\pi = 0^+$; $\nu_p = 2$ configuration $\pi^2 8^-(7/2^+[404] \otimes 9/2^-[514])$ in $K^\pi = 15^+$ ($\pi^2 8^- \otimes \nu^2 7^-$) band at 3653 keV; $\nu_p = 4$ configuration $\pi^4 11^-(7/2^+[404] \otimes 5/2^+[402] \otimes 9/2^-[514] \otimes 1/2^-[541])$ in $K^\pi = 18^+$ ($\pi^4 11^- \otimes \nu^2 7^-$) band at 4878 keV.[15,16]

The seniority ν-dependence of the pairing gaps have been investigated by the Lipkin-Nogami method in.[16] They showed that the bandhead pairing gap decreases approximately by $\Delta(\nu, \omega = 0) = (0.75)^{\nu/2}\Delta(\nu = 0, \omega = 0)$.

Figure 2 shows the PNC calculations of the proton pairing gaps for $\nu_p = 1, 3$ bands in ^{177}Ta and the gsb, and multiquasiparticle bands with $\nu_p = 2$ and 4 proton configurations in ^{178}W.[14,15] As a function of seniority ν, the pairing gap $\tilde{\Delta}(\nu)$ gradually decreases with increasing ν. The pairing gap reductions at bandhead ($\omega = 0$) calculated by PNC method are

$$\frac{\tilde{\Delta}_p(\nu = 1)}{\tilde{\Delta}_p(\nu = 0)} \approx 91\%, \quad \frac{\tilde{\Delta}_p(\nu = 2)}{\tilde{\Delta}_p(\nu = 0)} \approx 86\%,$$

$$\frac{\tilde{\Delta}_p(\nu = 3)}{\tilde{\Delta}_p(\nu = 0)} \approx 80\%, \quad \frac{\tilde{\Delta}_p(\nu = 4)}{\tilde{\Delta}_p(\nu = 0)} \approx 78\%,(3)$$

which are weaker than that given in Ref.[16]

4. Summary

The ω- and ν-dependence of the nuclear pairing gaps of multiquasiparticle bands in normally deformed nuclei are calculated using PNC method. PNC calculations show that the ω-dependence of pairing gaps $\tilde{\Delta}$ for the $\nu = 0$ ($\sim$qp-vacuum) bands is weaker than that predicted by the NHFB formalism. For the low-lying excited $\nu > 2$ ($\sim$multiquasiparticle) bands, $\tilde{\Delta}_p$'s and $\tilde{\Delta}_n$'s keep almost ω-independent. As a function of seniority ν, the bandhead pairing gaps $\tilde{\Delta}(\omega = 0, \nu)$, decrease slowly with increasing ν. Even for the highest seniority bands identified so far, the pairing gaps $\tilde{\Delta}_p(\omega = 0, \nu)$ and $\tilde{\Delta}_n(\omega = 0, \nu)$ remains larger than 70% of the bandhead value of the qp-vacuum band.

References

1. A. Bohr, B. R. Mottelson and D. Pines, *Phys. Rev.* **110**, 936 (1958).
2. A. Bohr and B. R. Mottelson, *Nuclear Structure, Nuclear Deformations, Vol. II* (Benjamin, New York, 1975).
3. A. B. Migdal, *Nucl. Phys.* **13**, 655 (1959).
4. S. T. Belyaev, *Mat. Fys. Medd. K. Dan. Vidensk. Selsk.* **31**, No. 11 (1959).
5. S. G. Nilsson and O. Prior, *Mat. Fys. Medd. K. Dan. Vidensk. Selsk.* **32**, No. 16 (1961).
6. J. Y. Zeng and T. S. Cheng, *Nucl. Phys. A* **405**, 1 (1983).
7. H. Molique and J. Dudek, *Phys. Rev. C* **56**, 1795 (1997).
8. B. R. Mottelson and J. G. Valatin, *Phys. Rev. Lett.* **5**, 511 (1960).
9. L. F. Canto, P. Ring and J. O. Rasmussen, *Phys. Lett. B* **161**, 21 (1985).
10. J. L. Egido, P. Ring, J. Iwasaki and H. J. Mang, *Phys. Lett. B* **154**, 1 (1985).
11. J. Y. Zeng, T. H. Jin and Z. J. Zhao, *Phys. Rev. C* **50**, 1388 (1994).
12. S. G. Nilsson, C. F. Tsang, A. Sobiczewski, Z. Szymaski, S. Wycech, C. Gustafson, I. L. Lamm, P. Möller and B. Nilsson, *Nucl. Phys. A* **131**, 1 (1969).
13. R. Bengtsson, S. Fraundorf and F. R. May, *At. Data. Nucl. Data. Tables* **35**, 15 (1986).
14. M. Dasgupta, G. D. Dracoulis, P. M. Walker, A. P. Byrne, T. Kibédi, F. G. Kondev, G. J. Lane and P. H. Regan, *Phys. Rev. C* **61**, 044321 (2000).
15. C. S. Purry, P. M. Walker, G. D. Dracoulis, T. Kibédi, F. G. Kondev, S. Bayer, A. M. Bruce, A. P. Byrne, W. Gelletly, P. H. Regan, C. Thwaites, O. Burglin and N. Rowley, *Nucl. Phys. A* **632**, 229 (1983).
16. G. Dracoulis, F. G. Kondev and P. M. Walker, *Phys. Lett. B* **419**, 7 (1998).

NUCLEON PAIR APPROXIMATION OF THE SHELL MODEL: A REVIEW AND PERSPECTIVE

Y. M. ZHAO*, Y. LEI, and H. JIANG

*Institute of Nuclear, Particle, Astrophysics and Cosmology,
Department of Physics, Shanghai Jiao Tong University, Shanghai 200240, China
*E-mail: ymzhao@sjtu.edu.cn
http://physics.sjtu.edu.cn/~ymzhao/*

We discuss the Nucleon Pair Approximation (NPA) of the shell model by a short review, which includes a history survey, its physical foundation, validity, and recent applications, as well as our perspectives.

Keywords: Pair approximation; transitional nuclei; shell model.

1. Introduction

The nuclear shell model (SM) provides a firm framework to study low-lying states in nuclei. However, the configuration space of the SM truncated to a single major shell for neutrons and protons is too huge to handle for medium-mass and heavy nuclei. In order to study properties of low-lying states, one must truncate the shell model space.

Pair approximation is among one of the practical approaches along this line. The wave functions of a given nucleus can be constructed in arbitrary basis of the assumed configuration spaces. They can be built by coupling the valence particles stepwise, and this can be easily realized via the well-known procedure of the Coefficients of Fractional Parentage (cfps). The configuration spaces can be also constructed by coupling the valence nucleon pairs with given spins. If all possible pairs are considered in the pair basis, the calculated results in the pair basis are equivalent to those of the exact shell model calculations. The Nucleon pair approximation (NPA) of the shell model refers to calculation in the nucleon pair basis with truncations of pairs, i.e., in the NPA only limited pairs are assumed to play important roles while others are neglected. The NPA is found to be very good and

---

*Corresponding author.

very useful approximations of the shell model for spherical and transitional regions.

Below we first present a brief discussion of the pair approximation,[5,6] including a short history survey, the framework, efforts by other groups and the orientations in our views.

2. History survey

The simplest pair approximation is the spin-zero pair (S for short) approximation for a single-j shell. Suppose the interaction between valence nucleons is the pairing interaction, the S pair approximation provides us with the exact solution for single-j shell. This scheme was generalized to many j shells (generalized seniority scheme[1]). Similarly, broken pair approximation was developed to study systems represented by configurations with only one or two pairs are non-S pairs. The S-pair approximation was generalized to include collective spin-two (D) pair, called SD-pair approximation.

The SD-pair approximation has been very popular during 1970's-1990's in piles of papers. This is mainly because of the great success of the the interacting boson model (IBM)[2] suggested by Arima and Iachello in the 1970's. The physics of the SD-pair approximation is well recognized. Attractive pairing interactions and quadrupople correlations are the most important in the residual interactions of the shell model Hamiltonian. These two interactions favors configurations constructed to a large extent by pairs with spin zero and spin two, i.e., S and D pairs. As an approximation one uses the SD pairs to construct the model space. For a comprehensive review, see Ref.[3] In the IBM, SD pairs are approximated to sd bosons.

There is a similar model, called the Fermion Dynamical Symmetry Model (FDSM).[4] The FDSM is similar to the IBM in the sense that it assumes the monopole and quadrupole correlations in truncating the model space, and applications of group theory. The advantages of the FDSM are its simplicity and the fermionic degree of freedom. The disadvantages of the FDSM are the specific features of both its nucleon pairs and the Hamiltonian. The overlaps between the FDSM wave functions and the exact shell model wave functions are usually very small.

In 1993 Chen developed the Wick theorem for coupled operators. Based on this technique, he proposed the nucleon pair approximation (NPA)[5] of the shell model in which one diagonalizes the shell model Hamiltonian in a coupled nucleon pair subspace. This technique was further developed for odd-A and doubly odd nuclei in Ref.[6] Important features of the NPA are that both the pair structures and configurations are flexible, and that the

Hamiltonian in the NPA is the same as that in the shell model. Explicit comparison between the NPA wave functions and the shell model wave functions demonstrates that the NPA is a very good approximation of the shell model.

3. Framework of the pair approximations

A collective nucleon pair with angular momentum r and z-component μ is defined by

$$A_\mu^{r\dagger}|0\rangle = \sum_{ab} y(abr)\left(C_a^\dagger \times C_b^\dagger\right)_\mu^r |0\rangle. \tag{1}$$

where a and b represent the quantum numbers nlj (often just j for simplicity) of an orbit in the valence shell. The quantities $y(abr)$ are called the structure coefficients of the collective pairs. The model space of pair approximations is constructed by coupling the pairs successively:

$$A_{M_N}^{J_N\dagger}(r_0 r_1 r_2 \cdots r_N, J_1 J_2 \cdots J_N)|0\rangle \equiv A_{M_N}^{J_N\dagger}|0\rangle$$
$$= \left(\cdots\left((A^{r_0\dagger} \times A^{r_1\dagger})^{J_1} \times A^{r_2\dagger}\right)^{J_2} \times \cdots \times A^{r_N\dagger}\right)_{M_N}^{J_N} |0\rangle, \tag{2}$$

where $A^{r_0\dagger} = C_a^\dagger$ for an odd valence nucleon number case, and $A^{r_0\dagger} = 1$ for an even valence nucleon number case.

The Hamiltonian of pair approximation is in principle the same as that in the exact shell model. However, for the sake of simplicity one usually takes a phenomenological form (which is easy to handle) of the shell model Hamiltonian as follows.

$$H = H_0 + V_0 + V_2 + H_Q. \tag{3}$$

Here H_0 is the single-particle energy term for valence neutrons and valence protons,

$$H_0 = \sum_{j\sigma} \epsilon_{j\sigma} C_{j\sigma}^\dagger C_{j\sigma},$$

where $\sigma = \pi, \nu$ correspond to the proton and neutron degrees of freedom, respectively. Also, V_0 and V_2 correspond to the monopole pairing and quadrupole pairing interactions, respectively,

$$V_0 = G_\pi^{(0)}\mathcal{P}_\pi^\dagger\mathcal{P}_\pi + G_\nu^{(0)}\mathcal{P}_\nu^\dagger\mathcal{P}_\nu,$$
$$V_2 = \sum_\sigma G_\sigma^{(2)}\mathcal{P}_\sigma^{(2)\dagger} \cdot \mathcal{P}_\sigma^{(2)}, \tag{4}$$

with

$$\mathcal{P}_\sigma^\dagger = \sum_{a_\sigma} \frac{\sqrt{2a_\sigma + 1}}{2}(C_{a_\sigma}^\dagger \times C_{a_\sigma}^\dagger)_0^0,$$

$$\mathcal{P}_\sigma^{(2)\dagger} = \sum_{a_\sigma b_\sigma} q(a_\sigma b_\sigma)\left(C_{a_\sigma}^\dagger \times C_{b_\sigma}^\dagger\right)_M^2,$$

The $q(ab)$ coefficients are given by $q(ab) = \frac{(-)^{j_a - 1/2}}{\sqrt{20\pi}}\sqrt{2j+1}\sqrt{2j'+1}$ $\langle nl|r^2|nl'\rangle C_{j1/2,j'-1/2}^{20}$, where $C_{a1/2,b-1/2}^{20}$ is the Clebsch-Gordan coefficient. Finally, H_Q is given by

$$H_Q = -\frac{1}{2}\chi(\eta_\pi Q_\pi + \eta_\nu Q_\nu)(\eta_\pi Q_\pi + \eta_\nu Q_\nu), \tag{5}$$

where

$$Q = \sum_{ab} q(ab)\left(C_a^\dagger \times \tilde{C}_b\right)_M^2 \tag{6}$$

is the quadrupole operator.

We note here that the Hamiltonian of the NPA is not restricted to the above form; it might assume other forms as well. For instance, Lei et al.[7] assumed the GXFP1A interactions in studying the validity of pair approximation for single-closed shell nuclei. The application of a general form of two-body interactions to heavy nuclei is now in progress.

4. Brief review of recent developments

In this section we discuss recent applications of the NPA, by Yoshinaga et al., and Luo et al., and ourselves.

The NPA was extensively applied to even-even nuclei with mass number around 130. This region has been of interest partly due to the O(6) pattern exhibited in low-lying states of a few even-even nuclei therein. Numerical calculations have been performed with different sets of parameters, see refs.[8–10] for details. These calculations paid attention to a few features of these nuclei. All calculations reasonably reproduce the O(6) behavior of the E2 transition rates as well as the energy levels. Some of the calculations concentrate on magnetic features. Generally speaking, the SD-pair approximations are good approximations for low-lying states, in particular, for the ground states, the first 2^+ and 4^+ states, and some non-yrast low-lying states. Yoshinaga et al. studied the contribution of a so-called H pair (alignment of two $h_{11/2}$ particles) in the yrast 6^+ and 8^+ states.

Recently the application of the NPA was extended to odd mass nuclei and odd-odd nuclei.[11–14] Here the main concerns are dominant configurations in low-lying states, as well as the energy level schemes, and E2 and M1 transition rates. The NPA is found to be very powerful in finding the dominant configurations of the low-lying states. For example, Xu $et\ al.$[12] showed that the $(\frac{1}{2})^-, (\frac{3}{2})^-, (\frac{5}{2})^-, (\frac{7}{2})^-$ and $(\frac{9}{2})^-$ states around 600keV of ^{209}Rn are essentially given by 2_1^+ state of the neighboring even-even core ^{210}Rn coupled with a single valence neutron hole in the $j = 5/2^-$ orbit. Similar phenomenon in ^{205}Pb and ^{207}Po. In Ref.,[13] Jiang $et\ al.$ demonstrated that low-lying states of some even-even, odd-mass and odd-odd nuclei with mass number around 200 are usually well represented by very simple configurations in collective nucleon-pair basis. In Ref.[14] Yoshinaga and Higashiyama studied the doublet bands of odd-odd Cs and La isotopes with mass number around 130, and claimed that these doublet bands are made of different angular momentum configurations of an unpaired neutron and an unpaired proton, weakly coupled with the quadurupole correlations of the corresponding even-even core.

The NPA was recently applied by Luo $et\ al.$ to study the phase transitions.[15] They showed that the NPA is able to reproduce various phase transitions if special hamiltonians are assumed.

5. Our perspectives

What physics does the NPA play the best? On the one hand, it is the most important to study how nuclei, with all its apparent complexity and diversity, can be constructed out of neutrons and protons, and their interactions; On the other hand, it is also extremely important to understand the features of complexity in nuclei as simple as possible. In the conventional shell model framework, this is very difficult, because the shell model space is usually too huge to figure out dominant configurations among billions of configurations in the model space, with a very small amplitude for each of them. In the NPA, however, extraction of dominant configurations is possible. In Ref.[13] wave functions of many low-lying states for nuclei with mass number around 200 are found to be well approximated by simple pairs.

There are two orientations of the NPA. One is to expand the configurations. Such expansions include the core excitations and dominant pairs with high spins. The core excitations were discussed for certain states. For heavy nuclei, excitations of cores are difficult to handle in conventional shell models but feasible in the NPA. Pair correlation in addition to SD pairs was already considered in many papers via a number of procedures. The other orientation of the NPA is to take more and more realistic interactions. Such efforts were devoted for single-closed nuclei. Hopefully we shall see the NPA calculations for heavy nuclei with both valence protons and valence neutrons in future.

The NPA was applied to spherical and transitional nuclei. Spherical nuclei can be also studied by using the shell model; in this case modern computers can perform the exact shell model calculations, because the dimension of model space is not too large. Transitional nuclei with large mass numbers are difficult for the shell model. The NPA is among the few microscopic approaches of low-lying states for such nuclei. There are many facets of collective motions in transitional nuclei. For sure the NPA will play an important role in studying such features.

There have been many discussions of the validity of pair approximation. Very recently, it becomes possible to perform the exact shell model calculations by using pair basis. This provides us with a powerful tool to investigate the wave functions of the NPA calculations in detail. This was done by Lei *et al.* in refs.,[7,16] where it was convincingly demonstrated that the pair approximation are indeed very good approximations of the full shell model space, although the dimensions of truncated pair basis are much smaller. However, there have been no convincing demonstration on validity of pair approximations yet, for heavy nuclei with both valence protons and valence neutrons. Prior to extensive applications of the NPA in the heavy nuclei, it will be extremely important to study the validity of the NPA by using the effective interactions.

6. Discussion and summary

In this paper we discuss the nucleon pair approximation of the nuclear shell model, including its history, physical foundation, validity, recent applications, and our perspective.

Finally we would like to point out, without details, that the pair approximations are very powerful not only in studying the structure of heavy nuclei, but also to exotic nuclear structure for light region. In the latter case many single-particle orbits are relevant in the configuration space.

Acknowledgements

We thank the National Natural Science Foundation of China for supporting this work under grants and 10975096 and 10675081. This work is also supported partly by Science & Technology Program of Shanghai Maritime

University under grant No. 20100086, and by Chinese Major State Basic Research Developing Program under Grant 2007CB815000. We also thank Prof. Arima for his constant encouragements in our works on the nucleon pair approximations.

References

1. I. Talmi, *Simple models of complex nuclei* (Harwood, New York, 1993).
2. F. Iachello and A. Arima, *the Interacting Boson Model* (Cambridge Univ. Press, England, 1987).
3. F. Iachello and I. Talmi, Rev. Mod. Phys. **59**, 339 (1987).
4. C. L. Wu, D. H. Feng, X. G. Chen, J. Q. Chen, and M. Guidry, Phys. Rev. **C36**, 1157 (1987).
5. J. Q. Chen, *Nucl. Phys. A* **626**, 686 (1997).
6. Y. M. Zhao, N. Yoshinaga, S. Yamaji, J. Q. Chen, and A. Arima, *Phys. Rev. C* **62**, 014304 (2000).
7. Y. Lei, Z. Y. Xu, Y. M. Zhao, and A. Arima, *Phys. Rev. C* **82**, 034303 (2010).
8. Y. M. Zhao, S. Yamaji, N. Yoshinaga, and A. Arima, *Phys. Rev. C* **62**, 014315 (2000); Y. M. Zhao, N. Yoshinaga, S. Yamaji, and A. Arima, *Phys. Rev. C* **62**, 024322 (2000); L. Y. Jia, H. Zhang and Y. M. Zhao, *Phys. Rev. C* **75**, 034307 (2007).
9. K. Higashiyama, N. Yoshinaga, and K. Tanabe, *Phys. Rev. C* **67**, 044305 (2003); N. Yoshinaga and K. Higashiyama, *Phys. Rev. C* **69**, 054309 (2004); T. Takahashi, N. Yoshinaga, and K. Higashiyama, *Phys. Rev. C* **71**, 014305 (2005).
10. Y. A. Luo and J. Q. Chen, *Phys. Rev. C* **58**, 589 (1998); Y. A. Luo, X. B. Zhang, F. Pan, P. Z. Ning, and J. P. Draayer, *Phys. Rev. C* **64**, 047302 (2001); Y. A. Luo, F. Pan, C. Bahri, and J. P. Draayer, *Phys. Rev. C* **71**, 044304 (2005); Y. A. Luo, F. Pan, T. Wang, P. Z. Ning, and J. P. Draayer, *Phys. Rev. C* **73**, 044323 (2006). X. F. Meng, F. R. Wang, Y. A. Luo, F. Pan, and J. P. Draayer, *Phys. Rev. C* **77**, 047304 (2008).
11. L. Y. Jia, H. Zhang and Y. M. Zhao, *Phys. Rev. C* **76**, 054305 (2007).
12. Z. Y. Xu, Y. Lei, Y. M. Zhao, S. W. Xu, Y. X. Xie, and A. Arima, *Phys. Rev. C* **79**, 054315 (2009).
13. H. Jiang, J. J. Shen, Y. M. Zhao, and A. Arima, preprint (to be published).
14. K. Higashiyama, N. Yoshinaga, and K. Tanabe, *Phys. Rev. C* **72**, 024315 (2005).
15. Y. A. Luo, Y. Zhang, X. F. Meng, F. Pan, and J. P. Draayer *Phys. Rev. C* **80**, 014311 (2009).
16. Y. Lei, Z. Y. Xu, Y. M. Zhao and A. Arima, *Phys. Rev. C* **80**, 064316 (2009); Y. M. Zhao, N. Yoshinaga, S. Yamaji, and A. Arima, *Phys. Rev. C* **62**, 14316 (2000).

BAND STRUCTURES IN DOUBLY ODD ^{126}I

Y. ZHENG, X. G. WU*, C. Y. HE, G. S. LI, X. HAO, L. L WANG, Y. LIU, X. Q. LI,

B. PAN, B. B. YU and L. WANG

China Institute of Atomic Energy, Beijing 102413, China
E-mail: wxg@ciae.ac.cn

L. H. ZHU

Department of Physics, Beihang University, Beijing 100191, China
Faculty of Science, ShenZhen University, Shenzhen 518060, China
E-mail: zhulh@buaa.edu.cn

The high-spin states in ^{126}I have been investigated by using in-beam γ-ray spectroscopy with the ^{124}Sn(^{7}Li, 5n)^{126}I reaction at a beam energy of 48 MeV. The previously known level scheme of ^{126}I has been extended and modified considerably by adding about 60 new γ-transitions and establishing 5 new bands. The backbendings in the yrast band 1 and the yrare band 3 are found both due to a pair of $h_{11/2}$ neutrons alignment. The configurations for the newly identified bands 2, 4, 5 and 6 have been assigned.

Keywords: Doubly odd nuclues; $\gamma - \gamma$ coincidence; high spin states.

The iodine nuclei with mass $A \sim 110 - 130$ have been of considerable interest as they lie in the transitional region between the primarily spherical $_{50}$Sn nuclei and the well-deformed $_{57}$La and $_{58}$Ce nuclei. They are expected to inherit the features from both the regions and exhibit a rich variety of nuclear structure led by significant variations of shapes and deformations with the configuration of the valance quasiparticles. Indeed, γ-ray spectroscopic investigations have revealed collective and noncollective structures representing diverse nuclear shapes in the odd-A $^{113-127}$I nuclei[1-7]. Previous studies have also observed collective bands based on the $\pi(d_{5/2}/g_{7/2}) \otimes \nu h_{11/2}$, $\pi g_{9/2}^{-1} \otimes \nu h_{11/2}$ and $\pi h_{11/2} \otimes \nu h_{11/2}$ quasiparticle configurations in the doubly odd $^{112-126}$I nuclei[8-16]. In this article, we report the new experimental results on high spin structures in ^{126}I.

Excited states in the doubly odd nucleus ^{126}I were populated via the heavy-ion fusion-evaporation reaction ^{124}Sn(^{7}Li,5n)^{126}I at a beam energy

252

of 48 MeV with beams provided by HI-13 tandem accelerator of China Institute of Atomic Energy. The target was an enriched self-supporting ^{124}Sn metallic foil of 4.6 mg/cm^2 thickness. The γ-rays emitted by the evaporation residues were detected using the multidetector array consisting of 12 BGO-Compton-suppressed HPGe detectors and 2 planar HPGe detectors. A total of $9.42 \times 10^7 \gamma - \gamma$ coincidence data was accumulated event by event. The recorded $\gamma - \gamma$ coincidence events were sorted into symmetry matrix and DCO matrix.

The level scheme of ^{126}I deduced from the present work is shown in Fig. 1. About 60 new γ-transitions have been observed and 5 new bands have been established. The placement of γ-transitions in the level scheme is based on their intensities, energy sums, and coincidence relationships. Spins of the level have been assigned on the basis of measured DCO ratios.

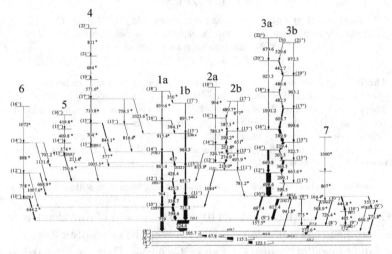

Fig. 1.   The level scheme of ^{126}I deduced from the present work. New γ transitions are indicated by asterisks.

The yrast states in band 1 have been established up to $I^\pi = 15^-$ level at 3.29 MeV and assigned to have $\pi g_{7/2} \otimes \nu h_{11/2}$ configuration [15]. In our work, we confirm the band and extend the band further up to the $I^\pi = 18^-$ level. The experimental Routhians e' and aligned angular momenta of bands 1-6 in ^{126}I are displayed in Fig. 2. The spin alignment for band 1 exhibit a gain of at least $9\hbar$ suggesting that the alignment of $h_{11/2}$ neutrons is involved. The total routhian surface (TRS) calculations for $\pi A \nu E$ show that the surface below the frequency of $\hbar\omega = 0.418$ MeV are very soft with respect to γ, but have a almost invariable shallow minimum with a near prolate shape having $\beta_2 \approx 0.14$ and $\gamma \approx -45°$. Cranked shell model (CSM) calculations were performed to make a configuration assignments of the bands 1 and 3. The deformation parameters were $\beta_2 = 0.101$, $\beta_4 = 0.002$ and $\gamma = +60°$. In the case of band 1, the calculations indicate alignment due to second pair of $h_{11/2}$ neutrons at a frequency of $\hbar\omega \approx 0.4$ MeV, which is in good agreement with experimental values (see Fig. 2). In view of these results and close similarity in systematics of neighboring odd-odd isotopes [12,13], previous assigned configuration for band 1 is adopted here as well. Band 1 has, therefore, been assigned to $\pi(d_{5/2}/g_{7/2}) \otimes \nu h_{11/2}^3$ configuration after the backbend based on the $\pi(d_{5/2}/g_{7/2}) \otimes \nu h_{11/2}$ configuration.

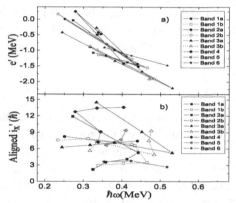

Fig. 2. Aligned angular momenta and experimental Routhians e' of bands 1-6. A reference core with Harris parameters $J_0 = 17$ $\hbar^2\text{MeV}^{-1}$ and $J_1 = 25$ $\hbar^4\text{MeV}^{-3}$ was used.

The levels in yrare positive-parity band 3 have been modified due to the identification of a new 665.6 keV transition. The ordering of the three $\Delta I=1$ 366.5 $\to$ 303.3 $\to$ 219.4 transitions reported in Ref.[16] turned out to be reversed by the present work. The previous configuration assignment $\pi h_{11/2} \otimes \nu h_{11/2}$ [16] for this band is adopted here as well. The TRS calculations for $\pi h_{11/2} \otimes \nu h_{11/2}$ configuration do show a energy minimum with a triaxial shape having $\beta_2 \approx 0.16$ and $\gamma \approx 30°$ at $\hbar\omega \approx 0.24$ MeV. Below this frequency two competing shallow minima are seen at both $\beta_2 \approx 0.15$, but with values of γ +50° and −50°. Therefore, the lowest states of the band 3 do not exhibit strong vibrational features, reveling an extremely γ-soft surface. Both the CSM calculations and the experimental data show that the crossing frequency in band 3 is $\hbar\omega \approx 0.45$ MeV. This crossing frequency and the gain in alignment (about 8 $\hbar$) in present data is consistent with $h_{11/2}$ neutrons.

Band 2 built on the 11^- level at 2213.7 keV was newly identified. We proposed that band 2 may be contain the same configuration of band 1 due to band 2 decay strongly into band 1. One can see from Fig. 2 that the spin alignment for band 2 is larger $4 - 5\hbar$ than band 1, which suggests the alignment of a pair of neutron in the $d_{5/2}/g_{7/2}$ orbital. Therefore, the configuration for band 2 could be assigned to $\pi(d_{5/2}/g_{7/2})\otimes\nu(d_{5/2}/g_{7/2})^2 h_{11/2}$.

We proposed that new bands 4 and 5 may be associated with $\pi(d_{5/2}/g_{7/2})\otimes\nu(d_{5/2}/g_{7/2})^2 h_{11/2}^3$ and $\pi(d_{5/2}/g_{7/2})\otimes\nu(d_{5/2}/g_{7/2})^2 h_{11/2}$ configurations, respectively. Besides, we suggested that another new band 6 should be associated with the $\pi(d_{5/2}/g_{7/2}) \otimes \nu h_{11/2}$ configuration coupled to the even-even Te core vibration.

In summary, the new level scheme of ^{126}I has been extended considerably by establishing 5 new bands and adding about 60 new γ transitions. The crossings in both the yrast band 1 and the yrare band 3 are attributed to a pair of $h_{11/2}$ neutrons. The configurations for the newly observed bands 2, 4, 5 and 6 have also been discussed.

Acknowledgments

The authors would thank to the crew of the HI-13 tandem accelerator in the China Institute of Atomic Energy for steady operation of the accelerator and for preparing the target. The authors also wish to thank Z. Y. Li and H. B. Ding for their help during experiment. This work is partially supported by the National Natural Science Foundation of China under Contract No. 10775184, 10675171, 10575133, 10575092, and 10375092 and by the Chinese Major State Basic Research Development Program through Grant No. 2007CB815000.

References

1. M. P. Waring et al., Phys. Rev. C **51**, 2427 (1995).
2. E. S. Paul et al., Phys. Rev. C **50**, 741 (1994).
3. M. P. Waring et al., Phys. Rev. C **48**, 2629 (1993).
4. Y. Liang et al., Phys. Rev. C **45**, 1041 (1992).
5. D. L. Balabanski et al., Phys. Rev. C **56**, 1629 (1997).
6. H. Sharma et al., Phys. Rev. C **63**, 014313 (2000).
7. Y. H. Zhang et al., HEP&NP, **26**,104 (2002).
8. E. S. Paul et al., J. Phys. G **21**,1001 (1995).
9. E. S. Paul et al., Phys. Rev. C **52**, 1691 (1995).
10. C. -B. Moon et al., Nucl. Phys. **A730**, 3 (2004).
11. C. -B. Moon et al., Nucl. Phys. **A728**, 350 (2003).
12. H. Kaur et al., Phys. Rev. C **55**, 512 (1997).
13. H. Kaur et al., Phys. Rev. C **55**, 2234 (1997).
14. C. -B. Moon, J. Korean Phys. Soc. **43**, S100 (2003).
15. R. J. Li et al., HEP&NP **29**,1 (2005).
16. C. -B. Moon, J. Korean Phys. Soc. **44**, 244 (2004).
17. J. Bured et al., Nucl. Phys. **A402**, 205 (1983).

LIFETIMES OF HIGH SPIN STATES IN ^{106}Ag

Y. ZHENG, X. G. WU*, C. Y. HE, G. S. LI, L. L WANG,

B. B. YU, S. H. YAO and B. ZHANG

China Institute of Atomic Energy, Beijing 102413, China
E-mail: wxg@ciae.ac.cn

L. H. ZHU

Department of Physics, Beihang University, Beijing 100191, China
Faculty of Science, ShenZhen University, Shenzhen 518060, China
E-mail: zhulh@buaa.edu.cn

Lifetimes of the high spin states in a pair of chiral candidate bands in ^{106}Ag were measured by means of the Doppler-shift attenuation method. The reduced transition probabilities $B(M1)$ and $B(E2)$ were deduced from these measurements. The $B(E2)$ values in both chiral candidate bands behave differently as well as $B(M1)$ values. In addition, the staggering of the $B(M1)$ values with spin was not observed. The experimental results are inconsistent with the ideal chiral predictions.

Keywords: Level lifetime; chirality; electromagnetic transition probabilities.

Since the first prediction of chirality in triaxial nuclei was pointed out by Frauendorf and Meng [1], candidate chiral partner bands with the configuration $\pi g_{9/2}^{-1} \otimes \nu h_{11/2}$ have been identified in ^{106}Mo [2], ^{100}Tc [3], $^{103-106}$Rh [4–9] and ^{105}Ag [10] isotopes in the $A \sim 100$ region. In ^{106}Ag, a pair of almost degenerate bands with negative parity has been reported in Ref. [11]. In another study [12], it was suggested that near the crossover point ($I \sim 14$) the two strongly coupled negative-parity rotational bands correspond to different shapes, which is different to the behavior expected from a pair of chiral bands. However, reduced $B(M1)$ and $B(E2)$ transition probabilities of the candidate chiral doublet bands were not discussed in Ref.[11,12]. Therefore, the measurement of $M1$ and $E2$ transition probabilities is crucial to settle whether the two chiral candidate bands in ^{106}Ag are in agreement with the ideal chiral predictions or not. In the present work, $B(M1)$ and $B(E2)$ transition strengths for the states in both chiral candidate bands in ^{106}Ag have been deduced from lifetimes measured using the Doppler-shift attenuation method (DSAM).

The high-spin states in the doubly odd nucleus ^{106}Ag were populated via the heavy-ion fusion-evaporation reaction ^{100}Mo(^{11}B,5n)^{106}Ag at a beam energy of 60 MeV provided by the HI-13 tandem accelerator of China Institute of Atomic Energy. The target consisted of a 1.56 mg/cm^2 thick ^{100}Mo backed on lead with a thickness of 8.03 mg/cm^2 in order to slow down and finally stop the recoiling nuclei. The deexciting γ-rays were detected using the multidetector array consisting of 10 BGO-Compton-suppressed HPGe detectors, 1 Clover detector and 2 planar detectors. A total of $2.3 \times 10^8 \gamma - \gamma$ coincidence events were accumulated in event by event mode. To obtain the line shapes of the investigated γ transitions in both chiral candidate bands of ^{106}Ag gates were set in the $\gamma - \gamma$ coincidence DCO matrices on the full peak shape of low-lying γ-ray transitions. The gated spectra were summed to obtain better statistic precision.

The lifetimes of the high spin states in ^{106}Ag are extracted by the analysis of the Doppler-broadened line shapes using the DSAMFT analysis program developed by Gascon [13]. The level scheme of ^{106}Ag relevant the present measurement is taken from Ref. [11]. The lifetimes of the levels from 11$^-$ to 15$^-$ in the yrast band and from 12$^-$ to 16$^-$ in the side band have been deduced. The representative examples of line shape fits along with the experimental spectra observed are shown in Fig. 1. The level lifetimes are the average values of results obtained from the forward angle and the backward angle. The extracted lifetimes and derived transition rates are listed in Table 1.

Table 1. Measured lifetimes of states in both chiral candidate bands, and corresponding $B(M1)$ and $B(E2)$ values. The uncertainty in stopping powers adds an 15% relative error on all the lifetimes.

band	I^π ($\hbar$)	E_γ (keV)	τ (ps)	$B(E2)$ (e^2b^2)	$B(M1)$ (μ_n^2)
yrast band	11$^-$	343.3	$1.50^{+0.59}_{-0.47}$	$0.059^{+0.027}_{-0.016}$	$0.73^{+0.34}_{-0.21}$
	12$^-$	833.7	$0.39^{+0.13}_{-0.12}$	$0.101^{+0.045}_{-0.025}$	$1.00^{+0.44}_{-0.25}$
	13$^-$	980.5	$0.68^{+0.31}_{-0.22}$	$0.050^{+0.024}_{-0.016}$	$0.44^{+0.21}_{-0.14}$
	14$^-$	552.2	$0.37^{+0.11}_{-0.10}$	$0.046^{+0.017}_{-0.010}$	$0.68^{+0.25}_{-0.15}$
	15$^-$	1146.3	$< 0.19^a$	> 0.074	> 0.94
side band	12$^-$	219.2	$0.75^{+0.88}_{-0.46}$	$0.312^{+0.495}_{-0.168}$	$6.69^{+10.62}_{-3.61}$
	13$^-$	269.9	$0.64^{+0.47}_{-0.33}$	$0.478^{+0.509}_{-0.202}$	$3.91^{+4.15}_{-1.65}$
	14$^-$	326.6	$0.52^{+0.57}_{-0.24}$	$0.308^{+0.264}_{-0.161}$	$2.63^{+2.26}_{-1.38}$
	15$^-$	429.5	$0.30^{+0.13}_{-0.11}$	$0.139^{+0.080}_{-0.042}$	$2.08^{+1.21}_{-0.63}$
	16$^-$	536.7	$< 0.18^a$	> 0.157	> 1.44

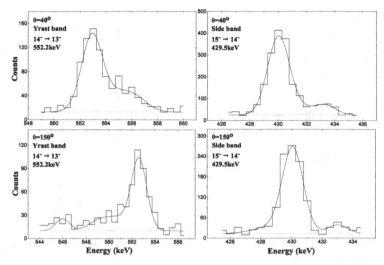

Fig. 1. (color online) Line shape fits for representative γ transitions observed at the forward (40°) and backward (150°) directions with respect to the beam axis in the two chiral candidate bands of ^{106}Ag.

Reduced $B(M1)$ and $B(E2)$ transition probabilities in the two chiral candidate bands in ^{106}Ag obtained from the present experiment are plotted versus spin I in Fig. 2. Within the experimental uncertainties, the $B(M1)$ values in both partner bands behave differently as well as $B(E2)$ values. In the angular-momentum region ($I^{\pi} = 12^{-} \rightarrow 15^{-}$) where the almost degeneracy of the energy levels of the two bands occurs, the experimental $B(E2)$ values in the side band are 2-10 times larger than those in the yrast band and the $B(E2)$ values become close with increasing spin. In the case of $M1$ transition, the experimental $B(M1)$ value in the side band is a factor of 6 to 7 larger than that in the side band at $I^{\pi} = 12^{-}$ and just differ by a factor about 2 at $I^{\pi} = 15^{-}$. In a word, the $B(M1)$ and $B(E2)$ values are rather different for the two partner bands, although they become close with increasing spin. Besides, the staggering of the $B(M1)$ values with odd and even spins was not observed. The experimental results are inconsistent with the ideal chiral predictions [5,14,15].

In summary, electromagnetic transition probabilities $B(M1)$ and $B(E2)$ in the two chiral candidate bands in ^{106}Ag are deduced from the lifetime measurements. The experimental $B(E2)$ values are larger in the side band than those in the yrast band as well as the $B(M1)$ values. In addition, the $B(M1)$ values are not staggering with odd and even spins. These characteristics are inconsistent with the interpretation of a composite chiral pair.

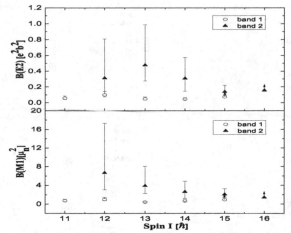

Fig. 2. Experimental determined $B(E2)$ and $B(M1)$ transition strengths in both chiral candidate bands in ^{106}Ag as a function of spin.

Acknowledgments

The authors would thank the crew of the HI-13 tandem accelerator in the China Institute of Atomic Energy for steady operation of the accelerator and for preparing the target. The authors also wish to thank C. Xu, J. G. Wang, and L. Gu for their help during experiment. This work is partially supported by the National Natural Science Foundation of China under Contract No. 10575133, 10675171, 10775184 and 10975191 and by the Chinese Major State Basic Research Development Program through Grant No. 2007CB815005.

References

1. S. Frauendorf and J. Meng, *Nucl. Phys.* **A617**, 131 (1997).
2. S.J. Zhu *et al.*, *Eur. Phys. J.* A **25** s01, 459 (2005).
3. P. Joshi *et al.*, *Eur. Phys. J.* A **24**, 23 (2005).
4. J. Timar *et al.*, *Phys. Rev.* C **73**, 011301(R) (2006).
5. C. Vaman *et al.*, *Phys. Rev. Lett.* **92** 032501 (2004).
6. T. Suzuki *et al.*, *Phys. Rev.* C **78** 031302(R) (2008).
7. J.A. Alcántara-Núñez *et al.*, *Phys. Rev.* C **69** 024317 (2004).
8. J. Timár *et al.*, *Phys. Lett.* **B598** 178 (2004).
9. P. Joshi *et al.*, *Phys. Lett.* **B595** 135 (2004).
10. J. Tima *et al.*, *Phys. Rev.* C **76** 024307 (2007).
11. C.Y. He *et al.*, *HEP&NP*, **30**, 166(S2) (2006).
12. P. Joshi *et al.*, *Phys. Rev. Lett.* **98**, 102501 (2007).
13. Gascon J *et al.* *Nucl. Phys.* **A513**, 344 (1990).
14. C. M. Petrache *et al.*, *Phys. Rev. Lett.* **96**, 112502 (2006).
15. S. Y. Wang *et al.*, *Chin. Phys. Lett.* **24**, 664 (2007).

EFFECT OF TENSOR INTERACTION ON THE SHELL STRUCTURE OF SUPERHEAVY NUCLEI

XIAN-RONG ZHOU* and CHEN QIU

*Department of Physics and Institute of Theoretical Physics and Astrophysics,
Xiamen University, Xiamen 361005, P. R. China*
E-mail: xrzhou@xmu.edu.cn

H. SAGAWA

*Center for Mathematics and Physics, University of Aizu,
Aizu-Wakamatsu, 965-8580 Fukushima, Japan*
E-mail: sagawa@u-aizu.ac.jp

In the frame of deformed Skyrme Hartree-Fock+BCS model, the effect of tensor interaction is discussed on the shell structure of superheavy nuclei. We compare the results of different Skyrme interactions; SLy5 without tensor interaction, and SLy5+T, T24 and T44 with tensor interaction. The large shell gaps of superheavy nuclei are found at Z=114 and Z=120 for protons and N=184 for neutrons at the spherical shape irrespective of the tensor correlations. It is also shown that the Z=114 and N=164 shell gaps are more pronounced by the tensor correlations in the case of SLy5+T interaction.

Keywords: Tensor interaction; superheavy nuclei; Skyrme Hartree-Fock.

1. Introduction

The prediction of the stability of superheavy nuclei is currently a topic of great interest in nuclear-structure physics. In superheavy nuclei, the spin-orbit interaction may induce substantial shell gaps due to the large density of single-particle energy levels, for instance, the Z=114 gap between the $2f_{7/2} - 2f_{5/2}$ spin-orbit partners as discussed in Ref.[1] Clearly the size of the Z=114 gap is determined by the strength of the spin-orbit splitting. The early calculations up to the late 1990s predicted that the nucleus of Z=114 and N=184 is the center of the island of long-lived superheavy elements (see, for example Ref.[2]), while more modern calculations with realistic effective nucleon-nucleon interactions suggest the next magic proton and neutron shells are at Z=120, N=172 or 184[3-5] and Z=124 or 126, N=184[2,6,7] (see

Refs.,[6,8-10] for comparisons of predictions of different Skyrme Hartree-Fock (SHF) and relativistic mean field (RMF) calculations). The differences in single-particle energies are responsible for the divergent predictions about magic numbers for superheavy nuclei.[11,12]

It is known that the tensor force[13] has the important effect on the spin-orbit splitting and will change the single particle energies and the shell structure of nuclei, although it had been neglected in mean field methods until very recently. A few years ago it was revived in Gogny[14] and Skyrme-type[15] models. Ref.[16] discussed the role of tensor terms of the Skyrme Hartree effective interaction on the spin-orbit splittings in the N=82 isotones and Z=50 isotopes. The experimental isospin dependence of these splittings can not be described by the Hartree-Fock calculations with standard Skyrme interactions, but is well reproduced by including the tensor interactions.[16] The similar improvements were realized in the study of f and p orbits in the nuclei around ^{48}Ca and ^{46}Ar.[17] In Ref.,[18] the effect of the tensor component of the Skyrme interaction on the single particle structure in superheavy elements is studied based on spherical Skyrme Hartree-Fock model.

In this paper, we study the effect of tensor interaction on the shell structure in superheavy nuclei using deformed SHF. We focus especially on the shell evolution of Z=114 in superheavy nuclei, where the tensor interaction may play an important role. This paper is organized as follows. A brief summary of the HF-BCS model with tensor interactions is given in Section 2. Section 3 is devoted to the study of shell structure of super-heavy nuclei. A summary is given in Section 4.

2. Formalism

We do the calculations using the deformed Skyrme Hartree-Fock (DSHF) calculation as in our previous works,[19-21] but taking into account the triplet-even and triplet-odd zero-range tensor interactions,

$$V_T = \frac{T}{2}\{[(\sigma_1 \cdot \mathbf{k}')(\sigma_2 \cdot \mathbf{k}') - \frac{1}{3}\mathbf{k}'^2(\sigma_1 \cdot \sigma_2)]$$

$$+ [(\sigma_1 \cdot \mathbf{k})(\sigma_2 \cdot \mathbf{k}) - \frac{1}{3}(\sigma_1 \cdot \sigma_2)\mathbf{k}^2]\}\delta(\mathbf{r}_1 - \mathbf{r}_2)$$

$$+ U\{(\sigma_1 \cdot \mathbf{k}')\delta(\mathbf{r}_1 - \mathbf{r}_2)(\sigma_2 \cdot \mathbf{k}) - \frac{1}{3}(\sigma_1 \cdot \sigma_2)[\mathbf{k}' \cdot \delta(\mathbf{r}_1 - \mathbf{r}_2)\mathbf{k}]\}. \quad (1)$$

With the contributions of tensor correlations, the spin-orbit potential is given by,

$$W_q = \frac{W_0}{2r}\left(2\frac{d\rho_q}{dr} + \frac{d\rho_q'}{dr}\right) + \left(\alpha\frac{J_q}{r} + \beta\frac{J_q'}{r}\right),\tag{2}$$

where the first term is from the Skyrme spin-orbit interaction whereas the second one includes both the central exchange and tensor contributions. Using the same notation as in Ref.,[16,17] we have

$$\alpha = \alpha_C + \alpha_T,\tag{3}$$

$$\beta = \beta_C + \beta_T,\tag{4}$$

$$\alpha_C = \frac{1}{8}(t_1 - t_2) - \frac{1}{8}(t_1 x_1 + t_2 x_2),\tag{5}$$

$$\beta_C = -\frac{1}{8}(t_1 x_1 + t_2 x_2),\tag{6}$$

$$\alpha_T = \frac{5}{12}U,\tag{7}$$

$$\beta_T = \frac{5}{24}(T+U),\tag{8}$$

where α_c and β_C are written in terms of standard Skyrme parameters, while α_T and β_T are related to the tensor interactions.

The contributions of central exchange and tensor terms to the SHF energy density are expressed as

$$\Delta H = \frac{1}{2}\alpha(J_n^2 + J_p^2) + \beta J_n J_p.\tag{9}$$

In the mean-field (particle-hole) channel we choose SLy5 parameterization without tensor interaction, and with tensor interactions, three parameter sets SLy5+T,[16] T24 and T44[22] are adopted. For SLy5+T, the tensor interactions are added on top of the existing parameter set SLy5 to describe the spin-orbit splittings of N=82 isotones and Z=50 isotopes. On the other hand, the interactions T24 and T44 are determined by the variational procedure for all the parameters including tensor terms.

A density-dependent surface delta interaction (DDDI)

$$V(\mathbf{r}_1, \mathbf{r}_2) = V_0'\left(1 - \frac{\rho(r)}{\rho_0}\right)\delta(\mathbf{r}_1 - \mathbf{r}_2),\tag{10}$$

is used for the pairing (particle-particle) channel, where $\rho(r)$ is the HF density at $\mathbf{r} = (\mathbf{r}_1 + \mathbf{r}_2)/2$ and $\rho_0 = 0.16$ fm^{-3}. The pairing strength is taken as $V'_p = -1462\,\mathrm{MeV\,fm^{-3}}$ and $V'_n = -1300\,\mathrm{MeV\,fm^{-3}}$ as in Ref.[23] A volume and surface mixed-type pairing interaction (MIX)

$$V(\boldsymbol{r}_1, \boldsymbol{r}_2) = V'_0 \left(1 - \frac{\rho(r)}{2\rho_0}\right) \delta(\boldsymbol{r}_1 - \boldsymbol{r}_2). \tag{11}$$

is also examined for the stability of superheavy elements.

3. Results

We perform the deformed SHF+BCS calculations, which include the tensor contributions to the energy density functional and the single particle potential. The parameters for tensor interaction employed in the present work are the same as Refs.[16,24] The total energies of $^{298}114_{184}$ are shown in Fig. 1 as a function of quadrupole deformation β_2 with SLy5, SLy5+T, T24 and T44 interactions, respectively. In order to check the dependence of results on the type of pairing interaction, we perform the DSHF+BCS calculations with the DDDI pairing (10) and also the MIX pairing. The results of DDDI and MIX pairings are given in the upper and lower panels of Fig. 1, respectively. We see two minima in the energy surface curves. i.e., at the spherical shape and at an excited prolate state with $\beta_2 \sim 0.6$ for all the interactions. While the two kinds of pairing give similar results, we notice that the barrier height between two minima is $1 \sim 2\mathrm{MeV}$ higher in the DDDI pairing than in the MIX pairing. The presence of tensor terms in the energy functional has an obvious impact on the energy surface. The SLy5+T interaction makes the spherical minimum deeper by about 1.5MeV than the higher local minimum at $\beta_2 \sim 0.6$. At the same time, the barrier height is also higher by about 1MeV in the case of SLy5+T. The other three interactions give the competing ground state and excited state with the energy difference of 0.78 MeV, 0.42 MeV and 0.12 MeV for SLy5, T24 and T44, respectively.

We plot the single particle energies of protons in $^{298}114_{184}$ in Fig. 2 as a function of deformation for four different Skyrme interactions. For all the interactions, we find $Z=126$ is the major shell gap in the spherical limit. For subshells, as discussed in the Ref.,[1] whether the $Z=114$ or 120 gap appears depending on the relative position of $2f_{5/2}$, $2f_{7/2}$ and $1i_{13/2}$ orbits. All the Skyrme calculations give the result in which $1i_{13/2}$ orbit lies between $2f_{5/2}$ and $2f_{7/2}$ orbits. We obtain competing $Z=114$ and $Z=120$ gaps at the spherical minimum for the SLy5, T24 and T44 interactions, while SLy5+T force gives a larger $Z=114$ gap (see also Table 1) . As was mentioned before, the occupation of $1i_{13/2}$ and $2f_{7/2}$ orbits enhances the

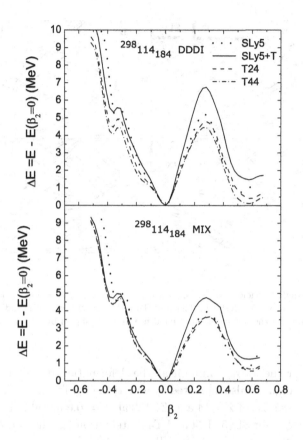

Fig. 1. The calculated total binding energy for ^{298}114 with DDDI (upper) and MIX (lower) pairing interactions as a function of deformation β_2. The energy is referred to the ground state energy ($\beta_2=0$) of each interaction.

spin-orbit splittings and makes the larger $Z=114$ shell gap than that of SLy5 interaction (without tensor interaction). On the other hand, for T24 and T44, the tensor effect is not clearly seen in the Nilsson diagram in Fig. 2. The spin-orbit interaction W_0 is taken larger in T24 and T44 interactions, i.e., $W_0=139.272$ and 161.367 MeV fm^{-5}, respectively, compared to $W_0=126$ MeV fm^{-5} for SLy5 and SLy5+T. Then α is taken positive for T24 and T44. Thus, in Fig. 2, the tensor effect of T24 and T44 interactions compensates with the larger W_0 strength and the net results give similar level schemes to that of SLy5. In the large prolate deformation side, it is interesting to notice that all the interactions give a clear energy gap at $Z=120$, which is

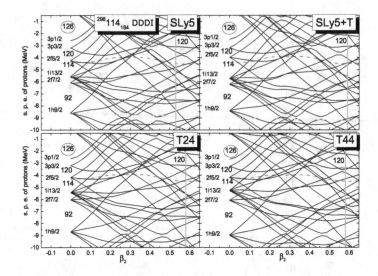

Fig. 2. The single-particle energies of protons in $^{298}114$ as a function of deformation parameter β_2 with SLy5, SLy5+T, T24 and T44. The Fermi energy is indicated by a long-dashed line and the deformation of local minimum is shown by a vertical gray line.

responsible for the local minimum at the deformation $\beta_2 \sim 0.6$ in Fig. 1 (the local minimum is also indicated by a vertical gray line in Fig. 2). We list the shell gap energies of $Z=114$ and $Z=120$ at $\beta_2 = 0.0$ in Table 1. From the table, we see that for SLy5, T24 and T44 interactions, both $Z=114$ and 120 gaps are competing, while for SLy5+T interaction, the energy gap at $Z=114$ is about two times larger than the $Z=120$ gap at the spherical minimum. It is also found in Ref.[18] that the inclusion of the tensor term leads to a small increase in the spin-orbit splitting between the proton $2f7/2$ and $2f5/2$ partners, opening the $Z=114$ shell gap over a wide range of nuclei. It should be also noticed that the $Z=126$ gap is always almost 2 times larger than those of $Z=114$ and 120.

The single particle energies of neutrons in $^{298}114$ with four different interactions are shown in Fig. 3. We obtain a large energy gap at $N=184$ for all the interactions. We can see also that the $N=164$ energy gap becomes much larger for SLy5+T because of the occupation of $1j15/2$ orbit and the negative α value, which make the spin-orbit splitting larger. In the large prolate deformation region, the $N=184$ gap appears at the deformation $\beta_2 \sim 0.6$ in the cases of T24, T44 and SLy5. We can see also the gap at $N=172$ for these three interactions. However, the deformed $N=184$ and

Table 1. The comparison of Z=114 and Z=120 energy gaps (in unit of MeV) at $\beta_2 = 0$ in $^{298}114$ for different Skyrme parameters SLy5, SLy5+T, T24 and T44. See the text for details.

Skyrme parameters	Z=114	Z=120	Z=126
SLy5	1.43	1.04	2.58
SLy5+T	1.77	0.80	2.64
T22	1.25	0.96	2.44
T44	1.11	0.89	2.17

N=172 gaps disappear for SLy5+T because $1i_{11/2}$ orbit becomes close to $2g_{9/2}$ and $1j_{15/2}$ orbits due to the tensor interactions and kills the energy gaps at $\beta_2 \sim 0.6$.

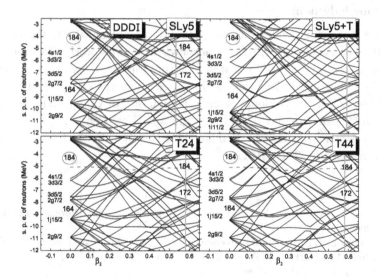

Fig. 3. Same as Fig. 2, but for the single particle energies of neutrons in $^{298}114$.

4. Summary

In order to study the effect of the tensor interaction on single-particle energies and shell structure in superheavy nuclei, we performed the deformed SHF+BCS calculations with four different Skyrme interactions SLy5, SLy5+T, T24 and T44. There is no tensor interaction in SLy5, while the tensors are included perturbatively in SLy5+T and, by the variational procedure in T24 and T44 parameter sets. At the spherical minimum we found the pronounced energy gaps at $Z=114$ and $Z=120$ irrespective of the tensor interaction. However, the tensor correlations of SLy5+T interaction make a larger shell gap at $Z=114$ than that at $Z=120$, which is the same as the spherical SHF results in Ref.[18] where pairing interaction is neglected. Near the deformed local minimum at $\beta_2 \sim 0.6$, we found again the $Z=120$ gap for all the four interactions. For neutron shells, we found the solid $N=184$ closure in all the four interactions and the tensor correlations of SLy5+T enhance the subshell closure of $N=164$ in the spherical shape. The large shell gap appears at $N=184$ even at the deformed local minimum of $\beta_2 \sim 0.6$ together with the $N=172$ gap in the cases of SLy5, T24 and T44. However the deformed gaps disappear by the tensor correlations of SLy5+T case.

Acknowledgments

Useful discussions with Gianluca Colò, K. Matsuyanagi, H. Hagino and A. P. Serveryukin are gratefully acknowledged. This work was supported by Japanese Ministry of Education, Culture, Sports, Science and Technology by Grant-in-Aid for Scientific Research under the program number (C (2)) 20540277, the National Science Foundation of China under contract Nos. 10605018 and 10975116, the Program for New Century Excellent Talents in University, and the Fundamental Research Funds for the Central Universities (No. 2010121011).

References

1. R.-D. Herzberg, P. T. Greenless, P. A. Butler, *et. al.*, *Nature* **442**, 05069 (2006).
2. S. Ćwiok, J. Dobaczewski, Heenen, P.-H., Magierski, P. W. Nazarewicz, Nucl. Phys. **A611**, 211 (1996); S. Cwiok, W. Nazarewicz, and P. H. Heenen, *Phys. Rev. Lett.* **83**, 1108 (1999).
3. K. Rutz, M. Bender, T. Bürvenich, T. Schilling, P.-G. Reinhard, J. A. Maruhn, and W. Greiner, *Phys. Rev.C* **56**, 238 (1997).
4. R. K. Gupta, S. K. Patra, and W. Greiner, *Mod. Phys. Lett. A* **12**, 1727 (1997).
5. S. K. Patra, C.-L. Wu, C. R. Praharaj, and R. K. Gupta, *Nucl. Phys. A* **651**, 117 (1999).
6. M. Bender, K. Rutz, P.-G. Reinhard, J. A. Maruhn, W. Greiner, *Phys. Rev. C* **60**, 034304 (1999).
7. A. T. Kruppa *et al.*, *Phys. Rev. C* **61**, 034313 (2000).
8. G.A. Lalazissis, M.M. Sharma, P. Ring, and Y.K. Gambhir, *Nucl. Phys. A* **608**, 202 (1996).

9. M. Rashdan, *Phys. Rev.* C**63**, 044303 (2001).

10. P.-G. Reinhard, M. Bender, and J. A. Maruhn, *Comments Mod. Phys., Part C* **2**, A177 (2002).

11. A. V. Afanasjev *et al.*, *Phys. Rev. C* **67**, 024309 (2003).

12. M. Bender *et al.*, *Nucl. Phys. A* **723**, 354 (2003).

13. Fl. Stancu, D. M. Brink and H. Flocard, *Phys. Lett. B* **68**, 108 (1977).

14. T. Otsuka, T. Matsuo, and D. Abe, *Phys. Rev. Lett.* **97**, 162501 (2006).

15. B. A. Brown, T. Duguet, T. Ostuka, D. Abe, and T. Suzuki, *Phys. Rev. C* **74**, 061303(R) (2006).

16. G. Colò, H. Sagawa, S. Fracasso, P. F. Bortignon, *Phys. Lett. B* **646**, 227-231 (2007).

17. Wei Zou, G. Colò, Z. Y. Ma, H. Sagawa, S. Fracasso and P. F. Bortignon, *Phys. Rev. C* **77**, 014314 (2008).

18. E. B. Suckling, P. D. Stevenson, *EPL* 90, 12001 (2010).

19. H. Sagawa, X. R. Zhou, Toshio Suzuki, and N. Yoshida, *Phys. Rev. C* **78**, 041304 (2008).

20. H. Sagawa, X. R. Zhou, and X. Z. Zhang, *Phys. Rev. C* **72**, 054311 (2005).

21. H. Sagawa, X. R. Zhou, X. Z. Zhang, and Toshio Suzuki, *Phys. Rev. C* **70**, 054316 (2004).

22. T. Lesinski, M. Bender, K. Bennaceur, T. Duguet, and J. Meyer, *Phys. Rev. C* **76**, 014312 (2007).

23. G. F. Bertsch, C. A. Bertulani, W. Nazarewicz, N. Schunck, and M. V. Stoitsov, *Phys. Rev. C* **79**, 034306 (2009).

24. M. Bender, K. Bennaceur, T. Duguet, P.-H. Heenen, T. Lesinski, J. Meyer, *Phys. Rev. C* **80**, 064302 (2009).

LIST OF PARTICIPANTS

Hong-Bo Bai (白洪波)
Department of Physics
Chifeng University
hbbai@vip.sina.com

Bao-Jun Cai (蔡宝军)
Department of Physics
Shanghai Jiao Tong University
landau1908@yahoo.cn

Chong-Hai Cai (蔡崇海)
Department of Physics
Nankai University
haicai@nankai.edu.cn

Xiang-Zhou Cai (蔡翔舟)
Shanghai Institute of Applied Physics
Chinese Academy of Sciences
caixz@sinap.ac.cn

Li-Gang Cao (曹李刚)
Institute of Modern Physics
Chinese Academy of Sciences
caolg@impcas.ac.cn

Lu Cao (曹璐)
School of Physical Science and
Technology, Southwest University
physicaolu@126.com

Wan-Cang Cao (曹万苍)
Department of Physics
Chifeng University

Xue-Peng Cao (曹雪朋)
China Institute of Atomic Energy
caoxp087@126.com

Zhong-Xin Cao (曹中鑫)
School of Physics
Peking University
zxcao@pku.edu.cn

Fei-Yan Chai (柴飞燕)
School of Nuclear Science and
Technology, Lanzhou University
chaify@lzu.edu.cn

Lie-Wen Chen (陈列文)
Department of Physics
Shanghai Jiao Tong University
lwchen@sjtu.edu.cn

Qi-Bo Chen (陈启博)
School of Physical Science and
Technology, Southwest University
chenqibo8681@126.com

Xiang-Rong Chen (陈向荣)
Department of Physics
Liaoning University
76803478.cici@163.com

Yong-Shou Chen (陈永寿)
China Institute of Atomic Energy
yschen@ciae.ac.cn

Peng-Cheng Chu (初鹏程)
Department of Physics
Shanghai Jiao Tong University
kyois@sjtu.edu.cn

Lian-Rong Dai (戴连荣)
Department of Physics
Liaoning Normal University
dailianrong@gmail.com

Guo-Xiang Dong (董国香)
School of Science
Huzhou Teachers College
daisy71411@126.com

Hong-Fei Dong (董鸿飞)
Department of Physics
Chifeng University
hongfeidong@126.com

Yong-Sheng Dong (董永胜)
Department of Physics
Jining Teachers College
dysh1973@163.com

Gong-Tao Fan (范功涛)
Shanghai Institute of Applied Physics
Chinese Academy of Sciences
fangongtao@sinap.ac.cn

De-Qing Fang (方德清)
Shanghai Institute of Applied Physics
Chinese Academy of Sciences
dqfang@sinap.ac.cn

Sheng-Qin Feng (冯笙琴)
College of Science
China Three Gorges University
fengsq@ctgu.edu.cn

Zhao-Qing Feng (冯兆庆)
Institute of Modern Physics
Chinese Academy of Sciences
fengzhq@impcas.ac.cn

Xi-Ming Fu (付熙明)
School of Physics
Peking University
fuximing123@sina.com

Yuan Gao (高远)
School of Nuclear Science and
Technology, Lanzhou University
gaoyuan@impcas.ac.cn

Zao-Chun Gao (高早春)
China Institute of Atomic Energy
zcgao@ciae.ac.cn

Long Gu (顾龙)
Department of Physics
Tsinghua University
gul06@mails.tsinghua.edu.cn

Xin Guan (关鑫)
Department of Physics
Liaoning University
ggguanxinnn@163.com

Jian-You Guo (郭建友)
School of Physics and Material
Science, Anhui University
jianyou@ahu.edu.cn

Shu-Qing Guo (郭树清)
School of Nuclear Science and
Technology, Lanzhou University
guoshuqing2009@lzu.edu.cn

Tao Guo (郭涛)
School of Physical Science and
Technology, Southwest University
Guotao19870@126.com

Rui Han (韩兹)
School of Physics
Peking University
cang.min@163.com

Rui Han (韩瑞)
School of Physical Science and
Technology, Southwest University
hanr367@swu.edu.cn

Hong-Jun Hao (郝红军)
School of Physics and Electrical
Engineering, Anyang Normal
University
Aysyhhj@sina.com

Chao He (何超)
School of Physics
Peking University
hechao@hep.pku.edu.cn

Chuang-Ye He (贺创业)
China Institute of Atomic Energy
chuangye.he@gmail.com

Xiao-Tao He (贺晓涛)
College of Material Science and
Technology, Nanjing University of
Aeronautics and Astronautics
hext@nuaa.edu.cn

Hui Hua (华辉)
School of Physics
Peking University
Hhua@hep.pku.edu.cn

Hai Huang (黄海)
School of Physics and Material
Science, Anhui University

Wei-Cheng Huang (黄维承)
Department of Physics
Tsinghua University
huangwc@mail.tsinghua.edu.cn

Juan-Xia Ji (姬娟霞)
School of Physical Science and
Technology, Southwest University
juanxia812@126.com

Hui-Ming Jia (贾会明)
China Institute of Atomic Energy
jiahm@ciae.ac.cn

Li Jia (贾麓)
Department of Physics
Liaoning Normal University
lulu19870524@126.com

Hui Jiang (姜慧)
Department of Physics
Shanghai Jiao Tong University
huijiang@shmtu.edu.cn

Wei-Zhou Jiang (蒋维洲)
Department of Physics
Dongnan University
wzjiang@seu.edu.cn

Chang-Feng Jiao (焦长峰)
School of Physics
Peking University
sillydream81@163.com

272

Gen-Ming Jin (靳根明)
Institute of Modern Physics
Chinese Academy of Sciences
jingm@impcas.ac.cn

Xiao-Kun Kang (康晓珅)
Department of Physics
Liaoning Normal University
kangxiaoshen05@163.com

Roberto Liotta
Royal Institute of Technology (KTH)
Alba Nova University Center
liotta@kth.se

Yang Lei (雷杨)
Department of Physics
Shanghai Jiao Tong University
Leiyang19850228@gmail.com

Ang Li (李昂)
Department of Physics
Xiamen University
liang@xmu.edu.cn

Cong-Bo Li (李聪博)
China Institute of Atomic Energy
lcbbch@sina.com

Fang-Qiong Li (李芳琼)
Guizhou University for Nationalities
lifq@sohu.com

Guang-Sheng Li (李广生)
China Institute of Atomic Energy
ligs@ciae.ac.cn

Hang Li (李杭)
Department of Physics
Liaoning Normal University
mumu_post@126.com

Jia-Xing Li (李加兴)
School of Physical Science and
Technology, Southwest University
lijx@swu.edu.cn

Jian Li (李剑)
School of Physics
Peking University
jlphyster@gmail.com

Li Li (李黎)
Department of Physics
Jilin University
Melodylili1986121@yahoo.com.cn

Lu-Lu Li (李璐璐)
School of Physics
Peking University
lilulu@itp.ac.c

Qing-Feng Li (李庆峰)
School of Science
Huzhou Teachers College
liqf@hutc.zj.cn

Shao-Xin Li (李绍歆)
Shanghai Institute of Applied Physics
Chinese Academy of Sciences
lishaoxin@sinap.ac.cn

Xiang-Qing Li (李湘庆)
School of Physics
Peking University
Lixq2002@hep.pku.edu.cn

Xiao-Wei Li (李晓伟)
Department of Physics
Chifeng University
nmglxw2000@126.com

Zhao-Xi Li (李兆玺)
School of Physical Science and
Technology, Southwest University
lzx516516@163.com

Zhi-Pan Li (李志攀)
School of Physics
Peking University
zpli.phy@gmail.com

Hao-Zhao Liang (梁豪兆)
School of Physics
Peking University
hzliang@pku.edu.cn

Chen Liu (刘晨)
School of Space Science and Physics
Shandong University at Weihai
chen880710@126.com

Gong-Ye Liu (刘弓冶)
Department of Physics
Jilin University
liugongye@126.com

Ling Liu (刘玲)
Department of Physics
Shenyang Normal University
Lling216@163.com

Xin Liu (刘鑫)
China Institute of Atomic Energy
liuxin@ciae.ac.cn

Yan-Xin Liu (刘艳鑫)
School of Science
Huzhou Teachers College
lyx1910@yahoo.cn

Gui-Lu Long (龙桂鲁)
Department of Physics
Tsinghua University
gllong@mail.tsinghua.edu.cn

Jing-Bin Lu (陆景彬)
Department of Physics
Jilin University
ljb@jlu.edu.cn

Hong-Feng Lu (吕洪凤)
College of Science
China Agricultural University
hongfeng@cau.edu.cn

Li-Jun Lu (吕立君)
Department of Physics
Chifeng University
lulijun0476@sina.com

Lin-Xia Lu (吕林霞)
Department of Physics
Nanyang Normal University
lvlinxia@sina.com

Lin-Hui Lu (吕林辉)
School of Physics
Peking University
lvlh@hep.pku.edu.cn

Yan-An Luo (罗延安)
Department of Physics
Nankai University
luoya@nankai.edu.cn

Chun-Wang Ma (马春旺)
Department of Physics
Henan Normal University
machunwang@126.com

Ke-Yan Ma (马克岩)
Department of Physics
Jilin University
ky525252@163.com

Yu-Gang Ma (马余刚)
Shanghai Institute of Applied Physics
Chinese Academy of Sciences
ygma@sinap.ac.cn

Hua Mei (梅花)
School of Physical Science and
Technology, Southwest University
meizi@swu.ed.cn

Jie Meng (孟杰)
School of Physics
Peking University
mengj@pku.edu.cn

Zhong-Ming Niu (牛中明)
School of Physics
Peking University
niuzhongming@163.com

Li Ou (欧立)
Department of Physics
Guangxi Normal University
only.ouli@gmail.com

Bin Qi (亓斌)
School of Space Science and Physics
Shandong University at Weihai
bqi@sdu.edu.cn

Chong Qi (亓冲)
School of Physics
Peking University
chongq@kth.se

Qiu Chen (邱晨)
Department of Physics
Xiamen University
48178332@qq.com

Cai-Wan Shen (沈彩万)
School of Science
Huzhou Teachers College
cwshen@hutc.zj.cn

Shui-Fa Shen (沈水法)
School of Nuclear Engineering and
Technology, East China Institute of
Technology
shenshuifa2006@163.com

Yue Shi (石跃)
School of Physics
Peking University
5531797rock@gmail.com

Bao-Xi Sun (孙宝玺)
Institute of Theoretical Physics
College of Applied Sciences
Beijing University of Technology
sunbx@bjut.edu.cn

Da-Peng Sun (孙大鹏)
School of Space Science and Physics
Shandong University at Weihai
sundapeng@sdu.edu.cn

Xiao-Yan Sun (孙小燕)
Shanghai Institute of Applied Physics
Chinese Academy of Sciences
sunxiaoyan@sinap.ac.cn

Qi Tan (谭奇)
Department of Physics
Liaoning Normal University
master1984@tom.com

Tmurbagen (特木尔巴根)
Department of Physics
Nanyang Normal University
tmurbagen@nynu.edu.cn

Wei-Xin Teng (滕维新)
Department of Physics
Liaoning Normal University
tengweixin911@yeah.net

Jun-Long Tian (田俊龙)
School of Physics and Electrical
Engineering
Anyang Normal University
tianjl@163.com

Yuan Tian (田源)
China Institute of Atomic Energy
tiany@ciae.ac.cn

Ya Tu (图雅)
Department of Physics
Shenyang Normal University
tuya_sy@sohu.com

Biao Wang (王彪)
School of Physical Science and
Technology, Southwest University
wbfxz@swu.edu.cn

En-Hong Wang (王恩宏)
School of Physics
Peking University
lionaria@163.com

Feng Wang (王枫)
School of Physics and Nuclear Energy
Engineering, Beihang University
nzma@163.com

Hai-Jun Wang (王海军)
Department of Physics
Jilin University
whj@jlu.edu.cn

Hong-Wei Wang (王宏伟)
Shanghai Institute of Applied Physics
Chinese Academy of Sciences
Wanghow@sinap.ac.cn

Jian-Guo Wang (王建国)
Department of Physics
Tsinghua University
wjg05@mails.tsinghua.edu.cn

Ning Wang (王宁)
Department of Physics
Guangxi Normal University
wangning@gxnu.edu.cn

Shou-Yu Wang (王守宇)
School of Space Science and Physics
Shandong University at Weihai
sywang@sdu.edu.cn

Si-Min Wang (王思敏)
School of Physics
Peking University
wangsimin@pku.edu.cn

Su-Fang Wang (王素芳)
Institute of Modern Physics
Chinese Academy of Sciences
wangsf@impcas.ac.cn

Ting-Tai Wang (王庭太)
School of Science
Zhongyuan University of Technology
tingtaiwang@163.com

Xiao-Bao Wang (王小保)
School of Physics
Peking University
wangxiaobaopku@yahoo.com.cn

Xin Wang (王欣)
Department of Physics
Shanghai Jiao Tong University
xinwang@sjtu.edu.cn

Yin Wang (王印)
Department of Physics
Chifeng University
wangyinwangyin@163.com

Yong-Jia Wang (王永佳)
School of Nuclear Science and
Technology, Lanzhou University
wangyjia05@lzu.cn

You-Bao Wang (王友宝)
China Institute of Atomic Energy
ybwang@ciae.ac.cn

Zong-Chang Wang (王宗昌)
Department of Physics
Nanyang Normal University
tmurbagen@yahoo.com.cn

Xiao-Guang Wu (吴晓光)
China Institute of Atomic Energy
wxg@ciae.ac.cn

Zhe-Ying Wu (吴哲英)
School of Chemical Engineering
Nanjing University of Science and
Technology
zheyingwu@gmail.com

Cheng-Jun Xia (夏铖君)
Institute of Theoretical Physics
College of Applied Sciences
Beijing University of Technology
newton_123@163.com

Jian Xiang (向剑)
College of Physical Science and
Technology, Southwest University
zhgxiangjian@163.com

Shao-Wu Xiao (肖绍武)
Department of Physics
Nanyang Normal University
xiaoshaowu@163.com

Ming-Xia Xie (解明霞)
Department of Physics
Liaoning Normal University
xiemingxia@sina.com

Chang-Jiang Xu (徐长江)
School of Space Science and Physics
Shandong University at Weihai
xueyutianma@126.com

Chuan Xu (徐川)
School of Physics
Peking University
xuchuan@hep.pku.edu.cn

Fu-Xin Xu (徐辅新)
Department of Physics
Anhui University
xufux@ahu.edu.cn

Xin-Xing Xu (徐新星)
China Institute of Atomic Energy
xuxinxing@ciae.ac.cn

Zheng-Yu Xu (徐正宇)
Department of Physics
Shanghai Jiao Tong University

Fu-Rong Xu (许甫荣)
School of Physics
Peking University
frxu@pku.edu.cn

Ting-Zhi Yan (颜廷志)
School of Energy Resources and
Power Engineering
Northeast Dianli University
ytz0110@163.com

Feng Yang (杨峰)
China Institute of Atomic Energy
martin@ciae.ac.cn

Dong Yang (杨东)
Department of Physics
Jilin University
dyang@jlu.edu.cn

Xing-Qiang Yang (杨兴强)
Department of Physics
Nanyang Normal University
yangxingqiang@yahoo.com.cn

Eing-Yee Yeoh (杨韵颐)
Department of Physics
Tsinghua University
yang-yy07@mails.tsinghua.edu.cn

Jiang-Ming Yao (尧江明)
School of Physical Science and
Technology, Southwest University
jmyao@swu.ed.cn

Shun-He Yao (姚顺和)
China Institute of Atomic Energy
xblll@sohu.com

Bei-Bei Yu (于蓓蓓)
China Institute of Atomic Energy
yubeibei0817@126.com

Shao-Ying Yu (于少英)
School of Science
Huzhou Teachers College
ysy@hutc.zj.cn

Yi-Fu Yu (于毅夫)
Department of Physics
Chifeng University

Xian-Qiao Yu (余先桥)
School of Physical Science and
Technology, Southwest University
yuxq@swu.edu.cn

Jin-Yan Zeng (曾谨言)
School of Physics
Peking University
jyzeng@pku.edu.cn

Chun-Li Zhang (张春莉)
School of Physics
Peking University
clzhang@pku.edu.cn

Da-Li Zhang (张大立)
School of Science
Huzhou Teachers College
zdl@hutc.zj.cn

Da-Peng Zhang (张大鹏)
Department of Physics
Fudan University
09300190033@fudan.edu.cn

Jia-Qi Zhang (张佳琦)
Department of Physics
Jilin University
9386592@163.com

Jin-Fu Zhang (张进富)
Department of Physics
Chifeng University
zhjinfu@sohu.com

Shuang-Quan Zhang (张双全)
School of Physics
Peking University
sqzhang@pku.edu.cn

Wei Zhang (张炜)
School of Physics
Peking University
philip@jcnp.org

Yu Zhang (张宇)
Department of Physics
Liaoning Normal University
dlzhangyu@yahoo.com.cn

Zhen Zhang (张振)
Department of Physics
Shanghai Jiao Tong University
mbimw@126.com

Zhen-Hua Zhang (张振华)
Institute of Theoretical Physics
Chinese Academy of Sciences
zhzhang@itp.ac.cn

En-Guang Zhao (赵恩广)
Institute of Theoretical Physics
Chinese Academy of Sciences
egzhao@mail.itp.ac.cn

Yu-Min Zhao (赵玉民)
Department of Physics
Shanghai Jiao Tong University
ymzhao@sjtu.edu.cn

Yun Zheng (郑云)
China Institute of Atomic Energy
zyxy_cloud@163.com

Pei Zhou (周培)
Shanghai Institute of Applied Physics
Chinese Academy of Sciences
zhoupei@sinap.ac.cn

Shan-Gui Zhou (周善贵)
Institute of Theoretical Physics
Chinese Academy of Sciences
sgzhou@itp.ac.cn

Xian-Rong Zhou (周先荣)
Department of Physics
Xiamen University
xrzhou@xmu.edu.cn

Sheng-Jiang Zhu (朱胜江)
Department of Physics
Tsinghua University
zhushj@mail.tsinghua.edu.cn